FERTILIZER SULFUR AND FOOD PRODUCTION

Fertilizer sulfur and food production

by

J.S. KANWAR
ICRISAT
Hyderabad
India

M.S. MUDAHAR
IFDC
Muscle Shoals, Alabama
USA

1986 Springer-Science+Business Media, B.V.

Library of Congress Cataloging in Publication Data

Kanwar, Jaswant Singh, 1922–
  Fertilizer sulfur and food production.

  "July 1985."
  Bibliography: p.
  1. Sulphur fertilizers--Tropics. 2. Food supply--
Tropics. 3. Sulphur deficiency diseases in plants--
Tropics. 4. Sulphur industry--Tropics. 5. Sulphur
fertilizers--Research--Tropics. 6. Sulphur fertilizers--
Government policy--Tropics. 7. Sulphur in nutrition.
I. Mudahar, Mohinder S. II. International Fertilizer
Development Center. III. Title.
S653.5.S84K36 1986   631.8'25   85-21614

ISBN 978-90-247-3244-9   ISBN 978-94-017-1540-9 (eBook)
DOI 10.1007/978-94-017-1540-9

Book Information

Paperback volume only available from the International Fertilizer Development Center, c/o Mr. Paul L.G. Vlek, P.O. Box 2040, Muscle Shoals, Alabama, 35662, USA, and from ICRISAT, Patancheru P.O., Andhra Pradesh 502 324, India.

Copyright

Table of Contents

VIII

List of Tables

List of Figures

Acronyms and Abbreviations

Fertilizers and Chemicals

Al	aluminum
AS	ammonium sulfate
B	boron
Ba	barium
C	carbon
Ca	calcium
CAN	calcium ammonium nitrate
$CaCl_2$	calcium chloride
$Ca(H_2PO_4)_2$	monocalcium phoshate
Cu	copper
DAP	diammonium phosphate
Fe	iron
FYM	farmyard mannure
H_2S	hydrogen sulfide
H_2SO_4	sulfuric acid
HOAc	acetic acid
K	potassium
K_2O	potassium, expressed as potassium oxide
KH_2PO_4	potassium dihydrogen phosphate
LiCl	lithium chloride
MAP	monoammonium phosphate
Mg	magnesium
Mn	manganese
Mo	molybdenum
MOP	muriate of potash (potassium chloride)
N	nitrogen
Na	sodium
NaH_2PO_4	sodium dihydrogen phosphate
$NaHCO_3$	sodium bicarbonate
NaOAc	sodium acetate
NH_4OAc	ammonium acetate
P	phosphorus
P_2O_5	phosphate, expressed as phosphorus pentoxide
PAPR	partially acidulated phosphate rock
S	sulfur
Se	selenium
SO_2	sulfur dioxide
SO_4^{-2}	sulfate ion
SSP	single superphosphate

TSP	triple superphosphate
Zn	zinc

Units of Measurements

Btu	British thermal unit
cm	centimeter
g	gram
GJ	gigajoule
ha	hectare
kcal	kilocalorie
kg	kilogram
km	kilometer
l	liter
m^3	cubic meter
mCi	millicurie
meq	milliequivalent
mg	milligram
ml	milliliter
mm	millimeter
mt	metric ton
ppm	parts per million
$	U.S. dollar
c.i.f.	cost, insurance, and freight
CPE	centrally planned economies
DME	developed market economies
DgME	developing market economies
f.i.o.	free in and out
f.o.b.	free on board

Organizations

BARI	Bangladesh Agricultural Research Institute
BRRI	Bangladesh Rice Research Institute
CIAT	Centro Internacional de Agricultura Tropical
FAO	Food and Agriculture Organization of the United Nations
FERTECON	Fertilizer Economic Studies, Ltd.
IARI	Indian Agricultural Research Institute
ICRISAT	International Crops Research Institute for the Semi-Arid Tropics
IFDC	International Fertilizer Development Center
IFPRI	International Food Policy Research Institute
IRRI	International Rice Research Institute

ODA	Overseas Development Administration
Sulexco	Sulfur Export Corporation
UNESCO	United Nations Educational, Scientific, and Cultural Organization
UNIDO	United Nations Industrial Development Organization
USDA	United States Department of Agriculture

Foreword

Fertilizer is a vital component of strategies for expanding foodproduction. The rapid growth in population and the widening food deficits inmany tropical countries of Asia, Africa, and Latin America call attention to those aspects of fertilization that have been neglected but are expected to yieldlarge economic payoffs in the future. Fertilizer sulfur falls into this category.

In the past fertilizer sulfur received little attention from researchers and policymakers since sulfur deficiency was not considered a serious problem. It was not a problem because of low crop yields, extensive cropping, and the incidental supply of sulfur through rain, irrigation water, manures, and sulfurcontaining fertilizers.

However, the situation has changed in the last three decades. Modernagriculture based on high crop yields, intensive cropping, improved crop varieties, and greater use of sulfur-free fertilizers and environmental regulations restricting sulfur emissions are creating large gaps between sulfur supply and sulfur requirements. Sulfur deficiencies are widespread and growing. Consequently, the full potential of a modern agricultural system in tropical countries is not being realized.

This research effort results from the recognition of the seriousness of the sulfur problem and its adverse impact on food production as well as IFDC's dedication to the development and transfer of economically efficient fertilizer technology to tropical countries. This study represents a comprehensive analysis of the technical and economic linkages between fertilizer sulfur and food production, and it provides guidelines for future directions in fertilizer sulfur research and public policy.

The project was jointly undertaken by Dr. J. S. Kanwar, Director of Research, International Crops Research Institute for the Semi-Arid Tropics (ICRISAT), and Dr. Mohinder S. Mudahar, Economist, International Fertilizer Development Center (IFDC). Dr. Kanwar, an eminent soil scientist, spent his 1982/83 sabbatical year at IFDC and participated in this research endeavor. This study is expected to provide needed impetus for national and international research and financial organizations to initiate and finance major fertilizer sulfur research and development programs.

The highlights of the study were published in 1983 as IFDC Technical Bulletin No. 27. It is hoped that this study will be of major significance to fertilizer researchers, extension agents, manufacturers, planners, and policymakers in their efforts to improve fertilizer use efficiency and alleviate world hunger.

Donald L. McCune
Managing Director
International Fertilizer Development Center

Acknowledgments

We are grateful to Dr. Donald L. McCune, Managing Director, IFDC, for inviting the first author to IFDC and suggesting that we study the problem of fertilizer sulfur and food production in relation to developing tropical countries of Asia, Africa, and Latin America; and to Dr. L. D. Swindale, Director General, ICRISAT, Hyderabad, India, for suggesting and encouraging the first author to spend his 1982/83 sabbatical year at IFDC.

We thank Dr. R. K. Cunningham of the Overseas Development Administration (ODA) of the United Kingdom for making available the research reports on multiyear sulfur programs in Nigeria and Kenya, conducted under an ODA assistance programby Mr. A. R. Bromfield and his associates. We have used these reports freelyand acknowledged them in the text. We are grateful to Dr. G. W. Cooke, Honorary Scientist, Rothamsted Experimental Station and formerly Chief Scientific Officer, Agricultural Research Council, London, for helping us establish the contact with this source of information and for his valuable comments on an earlier draft of this study.

We gratefully acknowledge the encouragement, support, and commentsreceived from IFDC management and Board members during the course of this study. We are grateful to Dr. Paul L.G. Vlek, Director, IFDC Agro-Economic Division, and Mr. Travis P. Hignett, IFDC Special Consultant to the Managing Director, who read the previous draft of the manuscript and made valuable comments. The assistance provided by Mrs. Susan Highfield at different stages of the study is also gratefully acknowledged. Finally, we are thankful to the Word Processing staff for typing several drafts of the manuscript and to the Communications staff for editing, developing artwork, and preparing the manuscript for publication.

J. S. Kanwar and
Mohinder S. Mudahar

1 Introduction

Statement of the Problem

The fast rate of growth in population, increasing food needs, and a widening gap between food consumption and production in the developing countries, particularly in the tropical countries of Asia, Africa, and Latin America, call attention to the need for research into those aspects of crop fertilization that have been neglected or that have the potential of affecting crop production significantly. An increase in land productivity and an expansion in the area under crops are two components of an economic strategy for increasing agricultural production. Fertilizer use is essential to this strategy because of its major contribution to crop yields, particularly in combination with irrigation and high-yielding crop varieties.

The importance of NPK fertilizers in increasing agricultural production is well recognized. The developing countries are making serious efforts to increase the use of fertilizers supplying these nutrients. However, sulfur (S), which is essential for synthesis of proteins, vitamins, and S-containing essential amino acids, has been ignored. In most cases, no more phosphorus (P) is required than S, but the use of P has received more attention, particularly in the tropics. The main reason for the lack of adequate attention to S in the developing tropical countries is that heretofore subsistence farming, low crop yields, and replenishment of S through the use of farmyard manure (FYM) and conventional S-containing fertilizers as well as irrigation water and addition of atmospheric S through rain have largely prevented S deficiencies.

The amount of S required for producing 1 mt/ha of cereal grain is about 4-5 kg/ha. Previously this amount was probably supplied through the sources mentioned above, and it could still meet the needs of subsistence farming except on soils that were inherently deficient in S or have become impoverished through losses of sulfate (SO_4) from leaching, immobilization, and crop removals. Moreover, at low levels of nitrogen (N) and P consumption the need for S was also low. As use levels increased, however, more S was also needed to ensure the high use efficiency of each nutrient. Sulfur deficiency was also masked somewhat by the acute deficiency of N and P. However, the situation is changing rapidly, and there is a growing awareness of S deficiency in many developed countries, particularly those where animal production is a major industry. Besides S deficiency, the acid rain in many developed countries is partly attributed to high sulfur dioxide (SO_2) content of atmosphere gases.

Classical examples of increased attention to S deficiencies include Australia, New Zealand, the United States, Canada, and Ireland, where the need for extensive use of S-supplying fertilizers is recognized in raising good pas-

1

tures and producing nutritive forages. In the United States S deficiency has been found in 37 states, not only in pastures but also in small-grain and coarse-grain cereals. Research and development programs on S use in crop production have become important. The number of scientists engaged in S research, the number of papers published, and the conferences held in the developed countries attest the importance of S fertilization for these countries.

The evidence of crop response to applid S in the developing countries of the tropics, the drastic decline in the additions of S because of the use of high-analysis and S-free fertilizers, and the growing problem of human hunger and malnutrition necessitate a critical look at the S fertilization problems in these regions. Sulfur in agriculture has many uses, but the four most important ones are these:

1. As a soil amendment for amelioration of saline alkali soils, calcareous soils, and soils of low permeability and for improving the quality of irrigation waters.
2. As a plant nutrient for correcting S deficiency, increasing crop yields, and improving the quality of crop produce.
3. As a chemical agent to acidulate phosphate rock and to manufacture phosphoric acid, phosphate fertilizers, ammonium sulfate (AS), and other S-containing fertilizers.
4. In pesticides, including fungicides.

The use of elemental S and S-containing substances such as gypsum and pyrites for soil reclamation purposes is well known. Sulfur will play a more significant role in improving productivity of lands that are saline, sodic, or calcareous or have low permeability. Sulfur also improves the poor-quality irrigation waters — a serious problem in semiarid tropics and arid regions. In these cases, however, the acidifying effect of S or the calcium (Ca) ion effect of gypsum is more important than the use of S as a nutrient. There are many publications by The Sulphur Institute in the United States, the Food and Agriculture Organization of the United Nations (FAO), the United Nations Educational, Scientific, and Cultural Organization (UNESCO), the British Sulphur Corporation, IFDC, and also other international and national organizations in developing countries that specifically discuss some of these aspects.

The use of S as a fertilizer nutrient, however, has not been adequately recognized in the developing countries of the tropics and subtropics. In view of the neglect from which S as a nutrient has suffered and the warning signals of S deficiency in the tropics and subtropics, this study is restricted to S as a plant nutrient and to S supply strategies.

Objectives

In this study attention is focused on the possible effects of S deficiency in soils and plants in the tropics and subtropics. A full understanding of this problem will lead to an appreciation of the need for S-related research as well as appropriate government policies.

It has not been possible to gain access to all the literature, published or unpublished, in different countries. The study aims at analyzing the available information to assess the nature, extent, and magnitude of the S problem and its relationship to food and nutritional needs of developing countries in the tropics and at formulating strategies for research on S fertilizers. It is hoped that this information will stimulate corrective action by national, regional, and international organizations and government policymakers.

The primary objective of this study is to analyze the economic importance of S in the fertilizer industry, food production, and the agricultural sector for the tropical countries. More specifically, the objectives are as follows:

1. To place in perspective the food and nutritional problems of tropical countries.
2. To examine the role of fertilizer S in designing strategies for food production.
3. To analyze the crucial role of S in plant nutrition and its effect on human and animal nutrition.
4. To examine the S status of tropical soils and its relation to crop production.
5. To analyze S deficiency, with particular reference to nature, causes, magnitude, loction, crops affected, and different diagnostic techniques.
6. To evaluate crop response to applied fertilizer S in the tropical countries of Asia, Africa, and Latin America.
7. To estimate aggregate S requirements, supplies, and implied S gaps in selected developing countries up to the year 2000.
8. To examine past performance, the current economic situation, and future outlook with respect to S demand, supply, prices, resources, and trade.
9. To evaluate alternative economic sources of fertilizer S and supply strategies, particularly in the context of indigenous S resources.
10. To examine the implications for research strategies and formulation of fertilizer policy with respect to fertilizer S, particularly in tropical countries.

In other words, the study deals with the agronomic, technological, economic, and policy aspects of fertilizer S in the context of tropical agriculture. It should serve as a basis for determining the appropriate role of fertilizer S in the design of alternative strategies and public policies for accelerating food production in developing countries.

2 Food and Nutrition Problems in Perspective

In order to design the research strategies and public policies needed to expand food production through appropriate use of fertilizer S, it is important not only to understand the nature and seriousness of the food problem but also to delineate the commodity sources of calories and protein and to examine the location of food and nutrition problems in developing countries.

Population and Food Production

Any realistic assessment of the world food problem must include an examination of various factors that influence the demand and supply of food. In this study, however, we limit the discussion of such factors to population – adeterminant of food consumption – as one side of the food equation and food production as the other. The regional distribution of world population and food production for 1981 is reported in Table 2.1. Developing countries as a group account for 74% of the world's population and yet produce only 47% of the total world food.[1] On the other hand, the developing market economies account for 50% of the world population and 29% of world food production.[2] The food problem is rather serious in developing market economies, particularly in Africa and the Far East. Consequently, a large number of people do not have access to adequate diets, and this seriously affects their life expectancy as well as work efficiency.

Furthermore, it is important to recognize three related issues. First, even though many developing countries have made major strides in food production, primarily as a result of the 'Green Revolution', their more recent performance (especially in Africa) has not been very impressive. Second, even though many developing countries are able to reduce the population growth rate, it is still too high in relation to growth in food production. Third, as reported in Table 2.1, a major share of cereals consumption in developed

1. The world's population will grow from 4.5 billion in 1981 to 6.35 billion in the year 2000. The rate of growth will slow only marginally. In absolute numbers, more people will be added annually in the year 2000 (100 million) than today (75 million). Approximately 90% of this growth will occur in the poorest countries of the world. The world population is expected to reach 10 billion by 2030 and will approach 30 billion by the end of the 21st century, according to the Council on Environmental Quality and the Department of State (1981).
2. Unless specified otherwise, in this study we have adopted FAO's classification of countries into economic classes and regions. According to this classification, the world is divided into three broad categories: (a) developed market economies (DME), (b) developing market economies (DgME), and (c) centrally planned economies (CPE). Developed countries consist of developed market economies, Eastern Europe, and the U.S.S.R., whereas developing countries consist of developing market economies and centrally planned countries from Asia, including China. Further details on this classification are given in Table 2.1.

5

6

Table 2.1. Regional distribution of world population and food production during 1981[a].

Region	Population (%)	Food production[b] (%)	Share of animal feed in regional cereal consumption[c] (%)
Developed market economies	17.7	36.4	75
North America	5.6	22.8	88
Western Europe	8.3	10.4	71
Oceania	0.4	1.4	61
Other developed	3.4	1.7	40
Developing market economies	49.8	29.3	12
Africa	8.6	3.8	5
Latin America	8.3	6.9	40
Near East	4.3	3.5	21
Far East	27.9	15.0	2
Centrally planned economies	32.6	34.3	41
Asian centrally planned	24.1	17.6	15
Eastern Europe + U.S.S.R.	8.4	16.7	67
Developed countries	26.1	53.1	72
Developing countries	73.9	46.9	13
World[d]	100 (4.5 billion)	100 (1.7 billion mt)	43

a. Derived from data reported in FAO (1982) and follows FAO regional classification.
b. Includes all cereals, pulses, root crops, and groundnuts. All the noncereals were converted into wheat equivalents based on calorie content. Rice refers to milled rice.
c. Consumption of cereal as food for people and feed for livestock. Derived from FAO (1977) and refers to 1972 – 74.
d. Totals are approximate due to rounding of data.

countries is in the form of feed for livestock to produce calories and protein for human consumption. Such a conversion process is generally not very efficient.[3]

Magnitude and Location of Food Deficits

Several national and international organizations, including FAO, International Food Policy Research Institute (IFPRI), U.S. Department of Agricul-

3. For further discussion and evidence on low conversion efficiency, see Balch and Cooke (1982) and Spedding, Walsingham, and Hoxey (1981).

ture (USDA), and the World Bank, periodically make projections for world food demand, supply, and food gaps. There may be, and often are, large differences in quantitative estimates of projected food gaps for a particular year and country across different organizations.[4] However, qualitatively all these estimates point to large projected deficits in developing countries, a large number of which do not even have the capability for commercial imports from food surplus countries. A large number of tropical countries fall into this category.

On the basis of trend projections, FAO (1979) has estimated that the net cereal deficit for 90 developing countries (excluding China) will be 91 million mt in 1990 and 153 million mt in the year 2000.[5] The share of individual regions in net cereal deficits during 2000 is estimated to be 29% in Africa, 33% in the Far East, 20% in the Near East, and 18% in Latin America. On the other hand, under a scenario of accelerated cereal growth, the net cereal balance for the same set of countries would be reduced to 52 million mt in 1990 and 88 million mt in the year 2000. The corresponding share of individual regions during the year 2000 is estimated to be 32% in Africa, 16% in the Far East, 45% in the Near East, and 7% in Latin America. In this scenario, the Indian subcontinent would change from a net deficit in cereals to a net surplus in cereals (4 million mt) by the year 2000, mainly in response to the realization of potential for accelerated production.

More recent estimates for food gaps (major staples as opposed to cereals) for selected countries and regions of developing market economies are reported in Table 2.2. The projected net food deficits for developing market economies are estimated to be approximately three times as great in the year 2000 as they were in 1977. The projeted deficits may increase or decrease, depending upon performance in the agricultural sector, national government policies with respect to food and nutrition, and population growth. The regions projected to face serious food problems are the Near East, west Africa, and upper (tropical) South America. The food gaps must be met through commercial imports and/or food aid. In this respect, these gaps have important implications for world food trade and food assistance. Otherwise, a large number of people may not be able to afford even the ba-

4. The differences in projected food deficits are generally due to (1) number of commodities included as food; (2) different initial conditions for food demand and supply; (3) differences in assumptions with respect to population growth, income growth, income elasticities, income distribution, land growth, multiple cropping, and farm technology; and (4) differences in functional forms of equations used in making projections.
5. Other studies dealing with world food problem and/or projections for food production, demand, and gaps include USDA (1974), Burki and Goering (1977), President's Science Advisory Committee (1967), Wortman and Cummings (1978), Crosson and Frederick (1977), IFPRI (1977), Paulino (1980) and Hopper (1981).

sic minimum diet.[6] Furthermore, food production in the year 2000 is projected to increase by 90% over that in 1977. This has important implications concerning the need for production incentives and the potential demand for modern inputs, including fertilizer.

Malnutrition in Developing Countries

National food self-sufficiency, an expressed national goal of many developing countries, does not always result in an adequate diet for every citizen. Inadequate nutrition may be attributed to several factors, including limited food supply, maldistribution of food, poverty, and lack of knowledge about nutrition. However, despite these and other socioeconomic factors, adequate domestic food production is an important component of strategies designed to eliminate the malnutrition problem.

The share and magnitude of population considered malnourished in selected countries and regions of developing market economies are reported in Table 2.3. Even if one does not agree with these estimates, the fact remains that large numbers of people do suffer and will continue to suffer from malnourishment. Furthermore, these results shed some light on the nature of food and nutrition problems. First, almost a quarter of the population (approximately one-half billion during 1981) in developing market economies is malnourished. According to FAO (1979) estimates, the share of the population below the critical 1.2 Basal Metabolic Rate in the year 2000 in developing market economies is expected to decline to 11% (387 million people) under a trend scenario and to 7% (242 million people) under an accelerated growth scenario. According to FAO(1977), however, the share of malnourished people in these countries has slightly increased from 24% in 1969-71 to 25% in 1972-74.

Second, almost one-half of the malnourished population is from South Asia, a region that is projected to have a food surplus during the year 2000. Third, even in grain-exporting countries such as Thailand, a significant share of the population is malnourished, which indicates that adequate production of food alone will not solve the malntrition problem. Fourth, even though there is no one-to-one relationship between calorie deficiency and protein deficiency, the large number of malnourished leads one to conclude that the population suffering from protein deficiency may be equally large and that the protein-energy-malnutrition syndrome prevails. Finally, as evidenced by FAO (1977, 1979), the malnutrition problem is much more

---

6. A report by the Council on Environmental Quality and the Department of State (1981) further reinforces the seriousness of the food problem. The report concludes that 'for hundreds of millions of the desperately poor, the outlook for food and other necessities of life will be no better. For many it will be worse' (Volume 1, p. 1).

Table 2.2. Estimated food production, consumption, and deficits in selected countries and regions of developing market economies[a].

Country/Region[b]	1977 (actual) (million mt)			2000 (projected)[c] (million mt)		
	Production	Consumption	Net deficit	Production	Consumption	Net deficit
India	134.0	122.7	11.3	234.3	220.8	13.5
Indonesia	24.9	27.4	-2.5	51.3	46.9	4.4
Philippines	8.9	9.4	-0.5	21.3	20.3	1.0
Sudan	3.9	3.5	0.4	9.9	7.5	2.4
Niger	1.9	1.7	0.2	1.5	3.0	-1.5
Nigeria	17.5	18.8	-1.3	19.4	41.0	-21.6
Kenya	3.9	3.0	0.9	13.4	8.0	5.4
Zimbabwe	1.8	1.9	-0.1	3.8	4.2	-0.4
Mexico	18.8	19.9	-1.1	46.1	46.9	-0.8
Brazil	40.6	42.1	-1.5	90.5	110.4	-19.9
Colombia	3.8	4.7	-0.9	9.7	10.4	-0.7
Asia	252.0	247.2	5.0	480.5	461.7	18.8
South Asia	167.5	158.5	9.0	306.3	291.1	15.2
East and Southeast Asia	84.6	88.6	-4.0	174.2	170.6	3.6
North Africa/Middle East	59.8	78.5	-18.7	119.8	177.1	-57.3
Africa group	16.7	28.2	-11.5	38.3	63.9	-25.6
Asia group	43.0	50.3	-7.3	81.5	113.2	-31.7
Sub-Sahara Africa	67.7	72.9	-5.2	112.7	149.1	-36.4
West Africa	31.1	33.5	-2.4	38.7	68.8	-30.1
Central Africa	11.6	13.0	-1.4	24.8	25.0	-0.2
East and South Africa	25.0	26.4	-1.4	49.3	55.3	-6.0
Latin America	103.5	103.2	0.3	226.6	232.0	-5.4
Central America/Caribbean	24.5	28.8	-4.3	58.5	64.3	-5.8
Upper South America	52.5	58.0	-5.5	114.2	144.2	-30.7
Lower South America	26.5	16.4	10.1	53.9	22.8	31.1
Developing market economies	726.2	753.7	-27.5	1 364.0	1 438.5	-74.5

a. Derived from preliminary estimates generated by IFPRI. Food refers to cereals, root crops, pulses, groundnuts, bananas, and plantain; noncereals are converted into cereal equivalents.

b. These individual countries were selected for a detailed analysis and to estimate S requirements and S gaps.

c. Projections for food production were based on 1961–77 trends for the commodities covered: projections for food consumption were based on 1966–77 trend for income growth and the United Nations population estimates.

serious among young children and women of child-bearing age, and this has serious implications with respect to their contribution to future economic growth.

Table 2.3. Share and magnitude of population considered undernourished in selected countries and regions of developing market economies.

Country/Region	Population with calorie intake below 1.2 MBMR[a]			
	% of total (1972/74)[b]	Million (1972/74)[b]	Million (1981)[c]	Million (2000)[c]
Bangladesh	38	27	34	54
Brazil	13	13	16	23
Colombia	28	7	7	11
Ethiopia	38	10	12	21
India	30	175	209	298
Indonesia	30	39	45	65
Kenya	30	4	5	11
Mali	49	3	4	6
Mexico	8	4	6	9
Niger	47	2	3	5
Pakistan	26	17	23	35
Philippines	35	15	18	27
Senegal	25	1	1	3
Sudan	30	5	6	10
Tanzania	35	5	6	13
Thailand	18	7	9	12
Zaire	44	10	13	22
Africa	22	68	86	149
Far East	27	286	340	490
Near East	11	19	24	36
Latin America	13	41	48	69
Developing market economies[d]	22	414	498	744

a. BMR refers to Basal Metabolic Rate, which is derived from basic physiological considerations. The coefficient, 1.2, was suggested by the FAO/World Health Organization (WHO) *ad hoc* Expert Committee on Nutrition and is determined by (1) an allowance for level of human activity (1.5 BMR) and (2) variation in BMR (0.8 BMR). This implies that 1.2 BMR = (1.5)(0.8) BMR. Thus, the critical food intake limit is 1.2 BMR and a person with food intake less than 1.2 BMR is likely to be undernourished.
b. Derived from FAO (1977). Regional estimates are from FAO (1979) and refer to 1975 and to 86 countries of developing market economies.
c. Calculated by assuming that the share of population below 1.2 BMR remains the same as it was during 1972–74. The population data for 1981 are from FAO (1982); for the year 2000 they are from the World Bank (1982). The regional World Bank classification is slightly different from that of FAO and does not include countries with population of less than 1 million.
d. Sum of four regions only. This does not include all the countries in developing market economies.

Sources of Calories and Protein Supply

Identification of the major food supply sources is a prerequisite to the design of research strategies and public policies directed at solving food problems in a particular country or region. This is especially important since food commodities produced locally determine the consumption patterns of the local population in that their tastes and preferences for food do not change suddenly. However, as per capita income improves, the consumers tend to shift their consumption patterns in favor of high-quality cereals, processed foods, fruits and vegetables, and livestock products.

The relative contribution of individual food commodities (expressed in wheat equivalents) to average per capita calorie and protein supply are reported in Tables 2.4 and 2.5, respectively. However, actual per capita supply should not be equated with actual per capita consumption of calories and proteins. The reasons for this distinction should be rather obvious. In comparing developing (low income, agricultural) countries with developed (high income, industrial) countries, two striking, but not surprising, conclusions emerge. First, the average per capita supply of calories and protein in developing countries is lower than that in developed countries, and the differences are rather substantial. The per capita calorie supply in developing countries was only 65% that of developed countries as a group. The corresponding share for protein supply was 58%. Second, vegetable products play a dominant role in developing countries by supplying 92% of calories and 79% of protein, as opposed to 68% of calories and 45% of protein in developed countries.

These results are instructive, but one must be very cautious in their interpretation and use for policy design. Large differences exist in consumption patterns across regions, countries, and even regions within a country. For example, cereals are an important source of calories and protein in different regions of developing market economies, but their relative contribution varies a great deal. While cereals account for almost two-thirds of the calories and protein available in the Near East and Far East, they account for one-half in Africa and two-fifths in Latin America. In Africa roots and tubers alone account for 21% of the calorie supply. Pulses are an important source of protein in all the developing market economies. In the Far East, pulses contribute almost as much protein as is contributed by meat, eggs, and milk combined. Finally, animal products in Latin America contributed 39% of the protein supply, which is almost two times their contribution in Africa and the Near East and 2.5 times their contribution in the Far East.

Table 2.4. Relative contribution of individual food items to average daily per capita calorie supply by region and economic group, 1972–74[a].

Food items	Developing market economies				Developed countries (%)	Developing countries (%)	World (%)
	Africa (%)	Latin America (%)	Near East (%)	Far East (%)			
Cereals	49	39	62	68	31	61	50
Roots and tubers	21	7	2	3	5	8	7
Sugar and honey	5	17	9	8	13	7	9
Pulses	4	5	3	4	1	4	3
Oilseeds and nuts	3	1	2	2	1	2	2
Meat, eggs, and milk	5	13	7	4	24	7	13
Vegetable products	94	84	90	94	68	92	83
Animal products	6	16	10	6	32	8	17
Calorie/capita (Kcal)[b]	2 114	2 538	2 443	2 044	3 378	2 212	2 548

a. Derived from data reported in FAO (1977), originally from FAO's food balance sheets. Per capita supply of each food item is obtained by dividing the quantity available for human consumption (production + imports − exports − fed to livestock − used as seed − used in the manufacturing sector − food loss ± changes in stocks) by the number of persons actually partaking of it and is expressed in terms of calories. However, average supply does not imply actual consumption per capita. The list of food items is not complete since only major food items are included in the table.
b. Average daily supply of kilocalories per capita.

Components of Food Production

It was pointed out earlier that local production is an important determinant of local consumption patterns. The relative contribution of individual commodities to food production during 1981 in developing market economies is analyzed in Table 2.6. The developing market economies account for 50% of world population but produce only 31% of all cereals and 40% of non-cereals.

The share of individual commodities in total food production also varies across regions. In the Far East, rice accounts for 51%, and it is a major staple in most of South Asia and Southeast Asia. In the Near East, wheat accounts for 65%. In Latin America, maize accounts for 49%. Finally, in Africa, sorghum and millet account for 28%, maize 25%, and root crops 23% of the region's total food production. The percentage share, without considering volume of production, could be misleading. For example, although pulses account for 4% in both Africa and the Far East, the total production of pulses in the Far East is more than four times that in Africa.

Table 2.5. Relative contribution of individual food items to average daily per capita calorie supply by region and economic group, 1972−74[a].

Food items	Developing market economies				Developed countries (%)	Developing countries (%)	World (%)
	Africa (%)	Latin America (%)	Near East (%)	Far East (%)			
Cereals	52	38	62	64	30	55	45
Roots and tubers	8	4	1	1	4	4	4
Pulses	11	12	6	10	2	10	7
Oilseeds and nuts	5	1	2	3	2	6	4
Meat, eggs, and milk	14	35	20	10	48	15	29
Vegetable products	81	61	79	85	45	79	65
Animal products	19	39	21	15	55	21	35
Protein/capita (grams)[b]	53	65	65	49	98	57	69

a. Derived from data reported in FAO (1977), originally from FAO's food balance sheets. Per capita supply of each food item is obtained by dividing the quantity available for human consumption by the number of persons actually partaking of it, and it is expressed in terms of protein. However, average supply does not imply actual consumption per capita. The list of food items is not complete since only major food items are included in the table.

b. Average daily supply of protein per capita.

It is important to know the relative contribution of different commodities in food production in order to assign priorities to strategies for expanding food production and crop research and for allocating farm inputs, including fertilizer. Since the main focus of this study is S, the following chapters will discuss the role of fertilizer S in accelerating food production in developing tropical countries. The use of fertilizer S influences both quantity and quality of crop production. Furthermore, the average S uptake varies across crops and crop groups that, in turn, determine the nature and magnitude of food production.

Table 2.6. Relative contribution of individual commodities to food production in developing market economies during 1981[a].

Major staple[b]	Production of food commodities (million mt)						Share of individual food commodities (%)						Share of DgME in world production (%)
	Africa	Latin America	Near East	Far East	DgME	World	Africa	Latin America	Near East	Far East	DgME	World	
Rice, milled	3.9	10.1	3.2	127.6	144.9	269.0	6	9	7	51	30	18	54
Wheat	4.3	15.0	31.5	49.5	100.4	458.2	7	13	65	20	21	31	22
Maize	15.5	55.8	4.9	19.9	96.1	451.7	25	49	10	8	20	31	21
Sorghum/millet	16.9	16.2	4.9	23.1	61.2	101.7	28	14	10	9	13	7	60
Root crops	14.0	11.4	1.6	15.1	42.2	132.5	23	10	3	6	9	9	32
Pulses	2.2	5.5	1.5	9.7	18.9	33.9	4	5	3	4	4	2	56
Groundnuts	4.2	0.9	1.0	8.0	14.1	20.3	7	1	2	3	3	1	69
Cereals[c]	40.6	97.1	44.5	220.1	402.6	1 280.6	67	85	92	87	84	87	31
Noncereals[d]	20.4	17.8	4.1	32.8	75.2	186.7	33	15	8	13	16	13	40
Total[e]	61.0	114.9	48.6	252.9	477.8	1 467.3	100	100	100	100	100	100	33

a. Derived from data reported in FAO (1982).
b. Production data were converted into wheat equivalents based on calorie content. The conversion ratios are 1.0:1.0 for milled rice, wheat, maize, sorghum, millet, and pulses; 1.0:0.25 for root crops; and 1.0:1.05 for groundnuts in shell.
c. Total of milled rice, wheat, maize, sorghum, and millet.
d. Total of root crops, pulses, and groundnuts.
e. Total of cereals and noncereals. Totals are approximate due to rounding of data.

3   Fertilizer Sulfur in Strategies for Food Production

The purpose of this chapter is threefold: (1) to briefly outline strategies for expanding food production, (2) to discuss the role of fertilizer in agricultural development, and (3) to examine the economic importance of fertilizer S in strategies for expanding food production and agricultural development in tropical countries of the world.

Strategies for Expanding Food Production

Despite the leading role of agriculture in economic development, governments of many low-income countries have not given the agricultural sector a central place in their development plans and public policies.[1] Consequently, governments of these countries must make stronger policy commitments in terms of resource allocation and economic incentives to initiate and sustain agricultural growth. Broadly, the strategies for expanding food production consist of (1) increasing the area under cultivation through land development and (2) increasing crop yields by intensive cultivation that uses multiple cropping, more and better fertilizers, and improved varieties.[2] Both strategies are important. However, the relative importance of these strategies depends on the country, the stage of its development, the resource endowments, the national goals, and its public policies.

Except possibly in parts of Africa and Latin America, extensive agricultural production (the increase in agricultural area) has limited scope. While marginal, low-fertility lands are being brought into cultivation, prime agricultural lands are being put aside for nonagricultural purposes. Some agricultural lands are also being lost through soil erosion, desertification, and deforestation. The development of new lands requires large capital investments, which many countries cannot afford. Increased investments and tightening land constraint will also increase the cost of food production.

The agricultural land per capita is declining in all the world regions, but at different rates. On the other hand, the amount of agricultural land per agricultural worker is increasing in developed countries (Figure 3.1). However, unlike the situation in Latin America, the amount of agricultural land per agricultural worker is gradually declining in Africa, Near East, Far East, and the countries with centrally planned economies in Asia, mainly China.

---

1. The role of agriculture in economic development is well documented. Among others, see Hayami and Ruttan (1971), Johnston and Kilby (1975), Johnston and Mellor (1961), Mellor (1965, 1976), and World Bank (1982).
2. Multiple cropping refers to growing more than one crop in a single year on the same piece of land.

This trend is expected to continue in the near future. According to FAO (1979), the share in projected agricultural production increase attributed to land expansion over the period of 1980-2000 is 54% in Latin America, 27% in Africa, 13% in Far East, 8% in Near East, and 28% in 90 developing countries (excluding China).

A large part of the additional food required to feed the growing population must come from an increase in land productivity. This will require both an increase in crop yields and multiple cropping. Undoubtedly, the contribution of the 'Green Revolution' (mainly in wheat and rice) to food production was very impressive. However, barring any unforeseen technological breakthroughs, similar increases in food production in tropical countries will require large capital investment and a concerted effort on the part of farmers, research workers, and policymakers.[3]

Regardless of the agricultural development strategy followed by a particular country, accelerating growth in food production will require the following: (1) massive investments in land development and the generation of irrigation capacity to relax land constraints, (2) the development of an agricultural research system and the necessary infrastructure to improve land productivity through higher yields and multiple cropping, (3) provision of economic incentives for the adoption of modern technology and expansion of food production, and (4) creation of capacity to produce, distribute, and use fertilizers to build and maintain soil fertility.

Fertilizer in Agricultural Development

Fertilizer contributes to economic development in many different ways. This includes the production of food and animal feeds, supply of energy through expanded production of energy crops, an increase in foreign exchange earnings, greater rural employment, and growth linkages through the industrialization and modernization of the agricultural sector. The available empirical evidence indicates that at least one-third of additional food production can be attributed to fertilizer use.[4]

---

3. The tropical regions have an advantage over the temperate regions in terms of abundance of sunlight energy and its impact on plant growth. However, as has been argued by Kamarck (1976), the climate has been a hindrance to economic development in the tropics. Some of the adverse factors encountered in the tropics are too much or too little rain; high temperatures; prevalence of pests and diseases for plants, animals, and humans; serious weed problem; and poor quality of soil resources. Kanwar (1982) has further echoed such concerns for soils and their impact on food production.

4. Some of the studies that deal with the contribution of fertilizer to food production are Bishop and Mudahar (1979), Christensen, Hendrix, and Stevens (1964), Free, Bond, and Nevins (1976), Herdt and Barker (1975), Mellor (1976), Pinstrup-Andersen (1976), Mudahar (1978), and von Peter (1980).

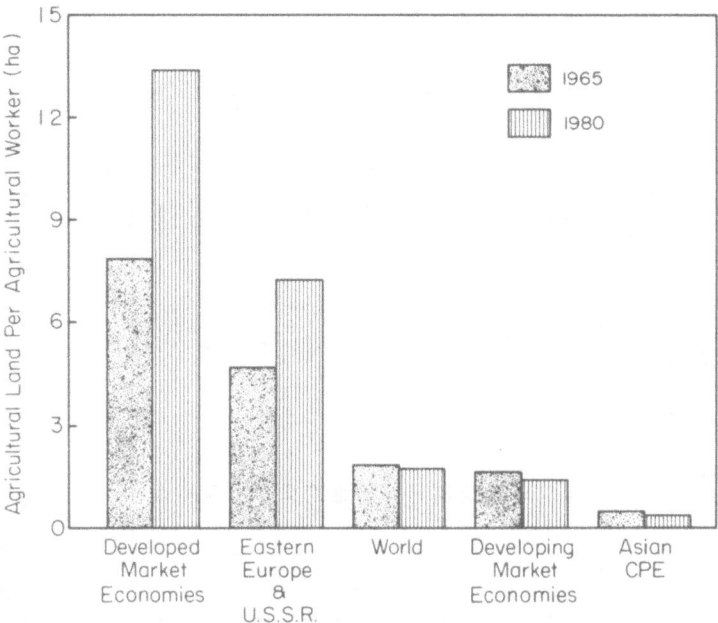

Figure 3.1. Regional Comparison of Agricultural Land per Agricultural Worker During 1965 and 1980.

According to the United Nations Industrial Development Organization (UNIDO) (1978) and FAO (1979), approximately 50%-70% of total fertilizer used in tropical developing countries is allocated to cereals. During 1982/83 the total consumption of nutrients (N + P_2O_5 + K_2O) in developing market economies was 22.8 million mt. If we assume that only 50% of fertilizer was used on cereal crops and that, on the average, the use of 1 mt of nutrient produces 10 mt of grain, then 114 million mt of grain can be attributed to fertilizer use alone. During 1982 the total cereal production (including rice as rice paddy) in developing market economies was approximately 490 million mt. As a result, it is estimated that 23% of total cereal production could have been the direct result of fertilizer use. This is a best first-order approximation; it is difficult to realistically isolate the contribution of fertilizer to cereal production because of the interactions among fertilizer, irrigation, fertilizer-responsive crop varieties, pesticides, and management.

The level, pattern, and growth in fertilizer consumption are reported in Table 3.1. The average fertilizer use (N + P_2O_5 + K_2O) in developing market economies is very low. During 1981 the average fertilizer use was 32 kg/ha, which was 15% of that in Western Europe, 27% of that in developed

Table 3.1. Level, pattern, and growth in fertilizer consumption for world regions in 1981[a].

Region	N (kg/ha)[f]	P_2O_5 (kg/ha)[f]	K_2O (kg/ha)[f]	Total[b] (kg/ha)[f]	N as % of total (%)	Growth rate[c] (%)	Share in world consumption Total[d] (%)	Incremental[e]
Developed market economies	56	32	30	118	47	2	38	31
North America	47	21	23	91	52	3	16	22
Western Europe	103	56	55	214	48	2	18	7
Oceania	6	26	6	38	16	1	1	1
Other DME	64	67	38	169	38	2	3	1
Developing market economies	19	9	4	32	59	8	20	33
Africa	4	4	2	10	40	7	1	3
Latin America	17	13	8	37	46	6	6	10
Near East	24	12	1	37	65	10	3	4
Far East	27	9	5	40	68	10	10	16
Centrally planned economies	65	31	24	119	55	7	42	35
Asian CPE	108	27	7	143	76	12	15	9
Eastern Europe & U.S.S.R.	47	32	30	110	43	5	27	27
Developed countries	53	32	30	115	46	3	65	58
Developing countries	32	12	5	48	67	9	35	42
World	41	21	16	79	52	4	100	100

a. Derived from FAO (1983), latest year for which fertilizer statistics are available.
b. Totals are approximate due to rounding of data.
c. Average annual compound growth rate in 1981 over 1969–71.
d. Total world consumption during 1982/83 was 114.7 million mt of nutrients ($N + P_2O_5 + K_2O$). Derived from data reported in FAO (1984).
e. From 1982/83 actual (FAO, 1984) to 1987/88 projected consumption (FAO/UNIDO/World Bank, 1983).
f. Kilograms per hectare of arable land and permanent crops.

market economies, 27% of that in centrally planned economies, and 41% of that in the world. The need to expand economically efficient fertilizer use cannot be overemphasized if we are to expand food production and reduce the probability of mass starvation.

The annual growth in average nutrient use from 1970 to 1981 in developing market economies has been rather impressive. The growth rate has been twice that of the world average and four times that of developed market economies. The recent fertilizer growth performance in developing market economies, though reassuring, could be misleading. Since the levels of fertilizer use in these countries are rather low, high growth rate reflects a relatively small increase in fertilizer use.

The average fertilizer use in developing countries is dominated by N. During 1981 N alone accounted for 67% of total nutrient use in developing countries, as opposed to 46% in developed countries. This reflects a certain degree of imbalance in fertilizer use since the uptake ratio of essential nutrients by crops is different from the corresponding nutrient supply ratio. There is no general rule about 'balanced' fertilizer supply since fertilizer requirements are specific to crops, soils, technology, and agroclimatic conditions. However, the areas with intensive cultivation are experiencing symptoms of deficiency in various essential nutrients, especially zinc (Zn) and S.

An adequate supply of nutrients is a prerequisite for expanded nd balanced fertilizer use. As discussed by Mudahar and Hignett (1982), the total nutrient consumption in the world during 1950/51 was 13.7 million mt. Only 8% of this was in developing countries, and the nutrients were used primarily on plantations and cash crops. During 1982/83 the world nutrient consumption was approximately 114.7 million mt, an increase of more than eightfold over 1950/51. The share of developing countries in world consumption increased to 35% during 1982/83.

According to FAO/UNIDO/World Bank (1983), the world nutrient consumption is projected to increase from 115 million mt in 1982/83 to 142 million mt in 1987/88. The share of developing market economies is expected to increase from 20% to 22% during this period. However, the developing market economies are expected to account for almost one-third of the incremental fertilizer consumption between 1982/83 and 1987/88.

The fertilizer requirements in the year 2000 are estimated to be 92.9 million mt in 90 developing countries (excluding China), according to FAO (1979), and 78.3 million mt in developing countries (excluding China), according to UNIDO (1978). These two projections are not really comparable; nevertheless, with a current fertilizer consumption in these countries of only 23 million mt, an almost fourfold increase in the next 20 years will pose a major challenge to policymakers. The implications of these projections are enormous for the finances and investments required for building fertilizer production and distribution capacity in these countries. Many countries may not have the necessary financial capability to procure fertilizer in order to meet their projected fertilizer (NPK only) requirements.

Importance of Sulfur

The economic importance of S is increasing rapidly as S deficiencies are becoming more widespread. The expanding role of S in accelerating food production in tropical countries is now being recognized. According to McCune (1982), 'Sulfur is so important in the tropics that, contrary to developed country practice, it must be treated as a major nutrient in the tailor-ing of fertilizers for tropical and subtropical agriculture.'[5] Sulfur consumption is considered one of the best indicators of economic progress. Sulfur and sulfuric acid (H_2SO_4) are widely used in the fertilizer, agricultural, and industrial sectors.

Sulfur is one of the essential plant nutrients, and it contributes to expan-

Table 3.2. Sulfur removed by selected crops in Brazil[a].

Crop	Yield, mt/ha	S removal kg/ha	S removal kg/mt[b]	S in residue as % of total S removal[c]	S removal in proportion to removal of other crop nutrients S	N	P	K	Ca	Mg
Maize	5.0	19	3.8	42	1.0	8.9	1.8	9.2	1.4	2.0
Rice paddy	4.0	9	2.3	44	1.0	9.3	1.6	9.9	2.3	1.0
Sorghum	2.5	7	2.8	43	1.0	9.3	1.4	6.9	2.3	1.7
Wheat	3.0	14	4.7	64	1.0	8.9	1.6	6.6	1.1	1.0
Barley	5.4	22	4.1	50	1.0	7.6	1.2	6.3	–	0.9
Soybeans	3.0	23	7.7	74	1.0	13.0	1.7	5.0	3.0	1.5
Field beans	1.0	25	25.0	60	1.0	4.1	0.4	3.7	2.2	0.7
Seed cotton	1.3	32	24.6	69	1.0	2.4	0.3	2.0	1.9	0.4
Peanuts	3.0	24	8.0	67	1.0	13.5	1.3	7.1	4.9	1.3
Sugarcane	100.0	12	0.1	–	1.0	11.0	0.7	9.2	1.1	1.6
Sugar beets	67.0	50	0.8	78	1.0	11.4	0.8	10.2	–	1.8
Coffee	2.0	27	13.5	89	1.0	9.4	0.7	8.6	5.3	1.2
Potatoes	40.0	11	0.3	73	1.0	18.2	0.7	20.0	4.7	1.5
Cassava	19.0	8	0.4	75	1.0	14.1	1.4	9.9	7.8	2.3

a. Calculated from data reported in Malavolta (1979).
b. Total S removal/mt of crop yield in the form of grains, seed cotton, beans, cane, or tubers. S removal for sugarcane, sugar beets, and potato is 0.12, 0.75, and 0.28 kg/mt, respectively.
c. Residue refers to above-ground residue and does not include roots.

5. Other studies that have recognized the vital role of S in food production in tropical countries include Coleman (1966), Fox (1980), 'Sulphur Classified as a Macronutrient' (1978), The Sulphur Institute (1975), and Terman (1978).

sion in crop yields in three different ways: (1) it provides a direct nutritive value, (2) it provides indirect nutritive value through improvements of calcareous and saline alkali soils, and (3) it improves the use efficiency of other essential plant nutrients, particularly N and P. As an illustration, S removal by selected crops in Brazil is reported in Table 3.2. The average removal of S varies from one crop to another and ranges from 10 to 50 kg/ha. A large share of S removed by a crop remains in the crop residues. Except for cereals, the amount of S removed by various crops is as great as or greater than the phosphorus (expressed as P rather than P_2O_5) removed.

The components of S supply and demand in the soil-plant-atmosphere system are developed in Figure 3.2. The relative importance of each of these components will vary from one system to another and will be discussed in the subsequent chapters. At this stage it will suffice to point out that S-

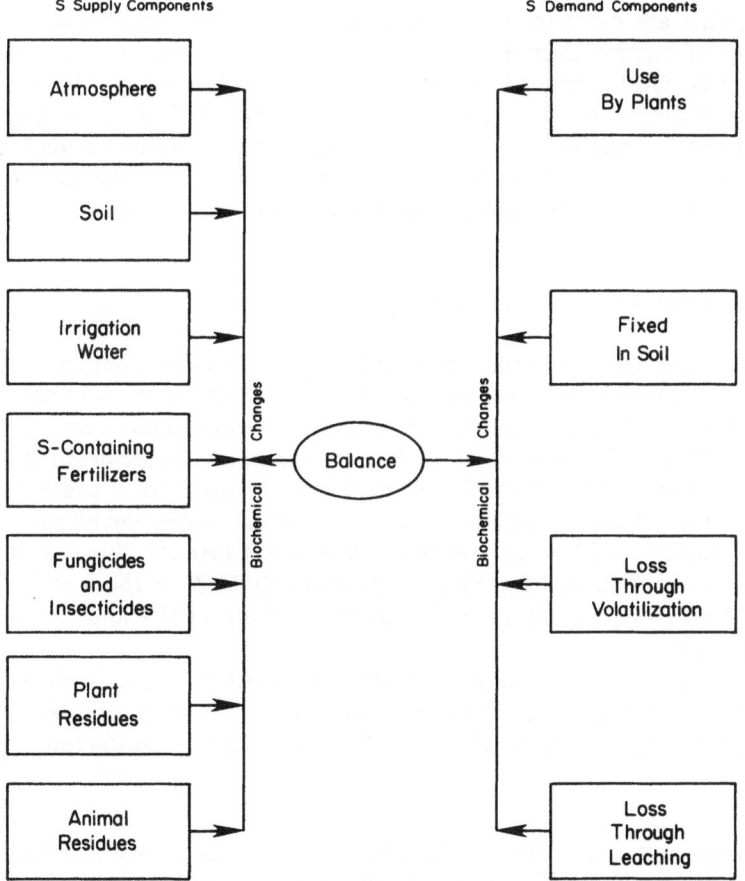

Figure 3.2. Components of Sulfur Supply and Demand in Soil-Plant-Atmosphere System.

containing fertilizer is only one source, albeit an important one, of S supplied to soils. Similarly, not all the S supplied to soil is taken up by the plant. A large share of it may be lost or may become fixed in the soil in compounds from which the S is not readily available. As with, use efficiency of S is rather low. Consequently, depending on the use efficiency, the amount of S that needs to be added to soil may be two to four times that of S removed by crops.

Sulfur plays an important role in protein synthesis and thus affects the quantity and quality of protein. It has been empirically established that for every 15 parts of N in protein there is 1 part of S, which implies that the N:S ratio is fixed within a very narrow range of 15:1. Clearly, a lack of S would reduce the amount of protein synthesized, even if there were plenty of N available to the plant. This relationship has important implications for human nutrition, especially in those countries where plant sources supply the bulk of the required proteins.

Proteins are essential for body growth. Fats and carbohydrates, good sources of energy, cannot be substituted for protein because they do not contain N. On the other hand, proteins are good but expensive sources of energy. Proteins supply the essential amino acids. Sulfur is an important constituent of methionine, one of the eight essential amino acids.[6] Sulfur is also required in the formation of chlorophyll and many other compounds that are involved in N fixation and photosynthesis by plants.

Sulfur in the Fertilizer Industry

Sulfur is also used to manufacture sulfuric acid, which is among the most versatile mineral acids. In the fertilizer industry sulfuric acid contributes in several ways. First, the manufacture of sulfuric acid produces usable energy − 1.32 GJ/mt of H_2SO_4 − in the form of steam and waste heat (Mudahar and Hignett, 1982). Second, sulfuric acid is used to manufacture S-containing fertilizers, including AS and single superphosphate (SSP). Third, sulfuric acid is used to manufacture wet-process phosphoric acid, which in turn is used to produce approximately 60% of the world's phosphate fertilizers, including triple superphosphate (TSP) and ammonium phosphates.

The role of S in the fertilizer industry and agricultural production is conceptualized in Figure 3.3. Primarily S is used to manufacture sulfuric acid. Approximately 0.33 mt of elemental S is required to manufacture 1 mt of

---

6. According to Passmore, Nicol, and Rao (1974), the eight essential amino acids include leucine, isoleucine, lysine, methionine, phenylalanine, threonine, tryptophan, and valine. In addition to these, histidine appears to be essential to the growth of infants.

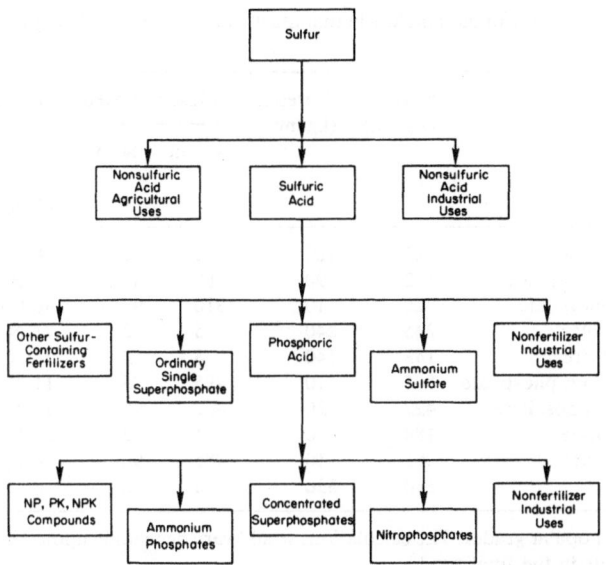

Figure 3.3. Role of Sulfur in Fertilizer Industry and Agricultural Production.

sulfuric acid. Some of the chemical processes in the fertilizer industry that directly or indirectly require sulfuric acid are described as follows:

Phosphate rock + sulfuric acid → single superphosphate

Phosphate rock + sulfuric acid →
wet-process phosphoric acid + phosphogypsum

Phosphate rock + phosphoric acid → concentrated superphosphates

Ammonia + phosphoric acid → ammonium phosphates

Ammonia + sulfuric acid → ammonium sulfate

Unfortunately, the fertilizer industry discards much of the S value of sulfuric acid in the form of byproduct phosphogypsum. As the fertilizer industry has shifted from low-analysis, S-containing fertilizers to high-analysis, S-free fertilizers, the amount of S wasted by the fertilizer industry has been rapidly increasing. The average S use and loss in manufacturing selected fertilizers are estimated in Table 3.3. The loss of S is highest in manufacturing wet-process phosphoric acid and those fertilizers that are derived from phosphoric acid. Hignett and Stangel (1982) have estimated that only about

Table 3.3. Average sulfur use and loss in manufacturing selected fertilizer products and intermediates[a].

Fertilizer material	S used (kg/mt)	S present (kg/mt)	S lost/removed		Nutrients in final product, %	
			kg/mt	% loss	N, P_2O_5, or K_2O	S
Single superphosphate	122	120	2	2	18 P_2O_5	12
Double superphosphate	172	90	82	48	25 P_2O_5	9
Triple superphosphate	320	10	310	97	46 P_2O_5	1
Ammonium sulfate	245	240	5	2	21 N	24
Sulfur-coated urea	143	140	3	2	38 N	14
Monoammonium phosphate	505	10	495	98	11 N + 55 P_2O_5	1
Diammonium phosphate	422	20	402	95	18 N + 46 P_2O_5	2
Potassium sulfate	184	180	4	2	50 K_2O	18
Phosphoric acid	495	10	485	98	54 P_2O_5	1
Sulfuric acid	330	330	0	0	0	33

a. Assuming popular grades. The calculations would vary depending upon the assumed nutrient contents in the final product.

10% of the S used in fertilizer manufacture appears in the finished products.

All those countries that produce wet-process phosphoric acid usually discard most of the byproduct phosphogypsum and hence lose the S contained in the phosphogypsum. In other words, one essential crop nutrient, S, is used and then discarded to produce another essential crop nutrient, P. For every metric ton of P_2O_5 in wet-process phosphoric acid, approximately 5 mt (4.62 mt, assuming no impurities) of phosphogypsum byproduct is produced. As far as S is concerned, for every metric ton of P_2O_5, 910 kg of S in the form of sulfuric acid (2.78 mt H_2SO_4 containing 32.7% S) is used; 860 kg of this ends up in phosphogypsum, which on the average contains 17% S (18.6% S, assuming no impurities).

The wet-process phosphoric acid capacity in the world during 1981 was approximately 29.5 million mt of P_2O_5. If all the existing capacity were fully used (which is not always the case), 26.8 million mt of S in the form of sulfuric acid would be required annually. Annual production of byproduct phosphogypsum would be 136 million mt, containing approximately 25 million mt of S (of 26.8 million mt used in the process). Most phosphogypsum is either stored in piles or ponds or discharged into rivers or oceans.

According to Weterings (1982), approximately 119 million mt of chemical gypsum (including 105 million mt of phosphogypsum) was produced in the world during 1981. Only about 16% was consumed (mainly in Japan, U.S.S.R., and Western Europe for building products), and the rest was either stored or discarded. According to Agarwal (1982), disposal of phosphogyp-sum is becoming a serious problem in India. In different parts of the world, phosphogypsum is being used (1) to produce AS, (2) as a soil amendment for saline soil, (3) as a source of S, (4) to manufacture building products such as blocks and plaster boards, and (5) to produce cement and sulfuric acid as coproducts. Appropriate use of phosphogypsum can result not only in economic benefits but also in reduced environmental problems, reduced disposal costs, and even a saving on foreign exchange for S-importing low-income countries.

The Sulfur Gap

The components of S supply and demand in the soil-plant-atmosphere sys-tem were conceptualized in Figure 3.2. The sources of S supply include atmosphere, soil organic matter, irrigation water, fungicides, plant residues, animal residues, and S-containing fertilizers. The relative importance of each of these supply sources varies with locality, level of industrialization, environmental considerations, and stage of economic development. On the other hand, S is taken up by the plant, fixed in the soil system, and lost through leaching or volatilization. Again, the relative importance of each of these pathways depends on soil, crops, and source of S.

In the last two decades, the developing tropical countries have exper-ienced several changes in agriculture and the fertilizer sector that have had a major impact on S availability and S requirements. With the introduction of the 'Green Revolution' technology, the aggregate requirements for S in-creased; this was mainly in response to an increase in crop yields and multi-ple cropping. Sulfur requirements will continue to increase rapidly in re-sponse to intensive cultivation (especially on marginal lands) in order to meet the ever-expanding demand for food and other agricultural products.

On the other hand, the aggregate S availability has been declining. First, most tropical countries never had (and many still do not have) any policy to supply S to soils. Sulfur was mainly supplied inadvertently through the use of those chemical fertilizers, such as AS and SSP, that also contained S.[7] Second, the fertilizer industry has been slowly replacing the S-containing fertilizers with S-free fertilizers, mainly because of high distribu-tion costs. SSP is being replaced by TSP, monoammonium phosphate (MAP), and diammonium phosphate (DAP); AS is being replaced by urea, ammonium nitrate, calcium ammonium nitrate (CAN), and DAP. Nothing is being done by the fertilizer industry or by governments to reverse the

---

7. AS and SSP are considered low-analysis fertilizers since no value is given to their S contents. If S is taken into account, just as N, P, and potassium (K) are, AS and SSP can be classified as high-analysis fertilizers.

trend, even in those areas facing S-deficiency problems. Third, the environmental concern for cleaner air has reduced the supply of S from the atmosphere.[8]

The net result of increased requirements and declining availability of S
is that the gap between them is widening over time. Consequently, S deficiency problems are becoming widespread. The low-income, food-deficit
tropical countries must bridge the S gap in order to realize their stated goals
of food selfsufficiency. These countries must formulate and implement fertilizer policies to adequately meet S requirements before the problem becomes too acute. The seriousness of the problem is illustrated by data from
India as a case study.

Aggregate nutrient ($N + P_2O_5 + K_2O$) consumption in India has increased from 66,000 mt in 1951/52 to 5.5 million mt in 1980/81 (FAI, 1983).
Not only has the consumption of the different nutrients been growing at
different rates over time (Figure 3.4), but the fertilizer sources of those
nutrients have also been changing (Figure 3.5). Annual consumption of S
has been estmated from the consumption of S-containing chemical fertilizers, including AS, ammonium sulfate nitrate, SSP, and potassium sulfate.

As a result, the estimated share of S in total nutrient ($N + P_2O_5 + K_2O
+ S$) consumption dropped from 54% in 1951/52 to 34% in 1965/66 and
to 5% in 1980/81. The case of the Indian Punjab, one of the progressive
agricultural states in India, is even more striking. In Punjab, the estimated
share of S in total nutrient ($N + P_2O_5 + K_2O + S$) consumption declined
from 10% in 1969/70 to 4% in 1974/75 and to 2% in 1979/80. The Punjab
Government has already taken corrective actions by advising farmers to use
gypsum as a source of S and even providing a price subsidy.[9] The fertilizer
S supply situation in many other developing tropical countries of the world
is not much different from that in India.

On the other hand, the amounts of S removed by crops, and hence S requirements, are increasing. Two main factors stand out. First, the agriculture industry is gradually shifting from subsistence to commercial agriculture, especially in those states experiencing the 'Green Revolution.' As a

8. Sulfur dioxide is emitted by the sulfuric acid plants and power plants burning hydrocarbon
fuels that contain S. Sulfur dioxide, through chemical reactions with air and water, is converted
into sulfuric acid. The sulfuric acid falls to the ground as what is commonly referred to as 'acid
rain.' Most developed countries have legislation restricting the emissions of SO_2. In most developing, nonindustrialized countries, this problem is not very serious since very little SO_2 is
emitted to the atmosphere anyway. The localities with industrial complexes, however, may face
these problems soon, if they do not already.

9. Subsidy on gypsum varies from 75% to small and marginal farmers up to a farm size of
3 ha to 50% for other farmers. During 1982/83 nutrient consumption per unit of gross
cropped area was 128 kg/ha, as opposed to only 37 kg/ha for India as a whole. Other Indian
states also have fertilizer subsidy programs, particularly for gypsum, lime, and other soil con-
ditioners. Formulation and effective implementation of selective subsidy programs can be used
to correct nutrient imbalance (Mudahar, 1978).

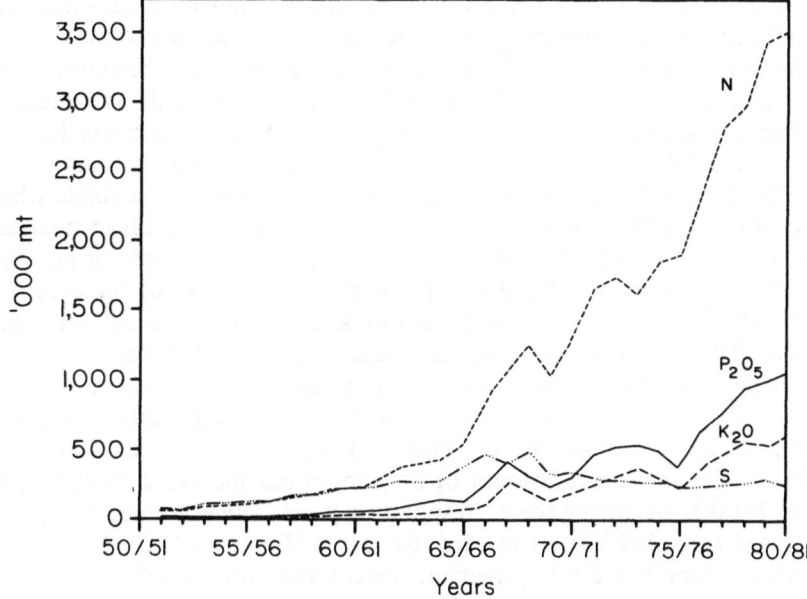

Figure 3.4. Growth in Distribution of Plant Nutrients in India.

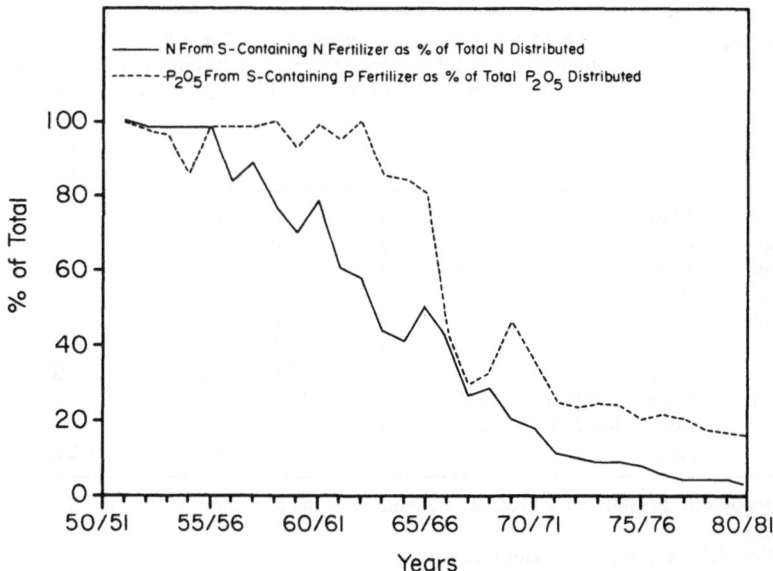

Figure 3.5. Evolution of Sulfur-Containing Nitrogen and Phosphate Fertilizer Use in Indian Agriculture.

result, more and more agricultural products are sold to urban-industrial complexes. Sulfur removed by these products is not recycled to the field, as is the case in subsistence agriculture. Second, agriculture is becoming more modern and intensive, and this includes higher crop yields and multiple cropping. The impact of changing technology on S requirements is illustrated in Table 3.4. On the average, the amount of S removed increases from 6 kg/ha/year to 20 kg/ha/year as the farmer switches from a single wheat crop using traditional technology to a double crop (in this case wheat and rice rotation so popular in Punjab) using modern technology. This represents an increase of 333%. The percentage increase of removal over that with a single rice crop using traditional technology is even more spectacular. These changes are gradually occurring all over India.

The declining S availability and the increasing S requirements are widening the S ga. In India the incidental supply of S through fertilizers is estimated to have declined from 395,000 mt in 1965/66 to 250,000 mt in 1980/81. The estimated amount of S removed has increased from 524,000 mt to 784,000 mt during the corresponding years. The estimated S gap thus increased from 129,000 mt in 1965/66 to 534,000 mt in 1980/81. The gap between S supply and S requirements (accounting for use efficiency of ap-

Table 3.4. Impact of changing technology on average annual sulfur requirements in tropical agriculture: A case of wheat and rice cropping system[a].

Cropping system	Crops grown	Technology regime	Average yield[b] (mt/ha)	S removal[c] kg/mt	kg/ha	Total S requirements[d] (kg/ha/year)	% increase in S requirements over Local wheat	Local rice
Single	Wheat	Local	1.5	4	6.0	10.5	–	
Single	Rice	Local	1.2	3	3.6	6.3		–
Single	Wheat	Modern	3.0	4	12.0	21.0	100	
Single	Rice	Modern	2.8	3	8.4	14.7		130
Double	Wheat	Local	1.5	4	6.0			
	Rice	Local	1.2	3	3.6			
	Total				9.6	16.8	60	167
Double	Wheat	Modern	3.0	4	12.0			
	Rice	Modern	2.8	3	8.4			
	Total				20.4	35.7	240	467

a. Hypothetical system which realistically simulates the condition of Punjab, India, agriculture.
b. Rice yield is in terms of paddy rice.
c. Removal of S by both grain and crop residue.
d. 1.75 times the S removed, implying 57.1% use efficiency. Reducing use efficiency to one-half could double the corresponding S requirements.

plied S) has increased much more.[10] In principle, the case of India demonstrates the nature and seriousness of S problems in tropical agriculture. Clearly these widening S gaps cannot be bridged by chemical fertilizers alone, but the pressure is building on the policymakers to recognize the S problem and to design realistic national S supply strategies.

Social Cost of Inadequate Sulfur Use

Often people do not realize that fertilizer policy, or the lack thereof, generates both social benefits and social costs that are generally shared differently by different segments of the society. Discussion in this section will be limited to the loss in agricultural output due to lack of S fertilizer.

The average response of rice paddy to different levels of S in the form of AS is reported in Table 3.5. The results are based on experiments conducted in the South Sulawesi province of Indonesia. First, an increase in rice yield over control was 1.59 mt/ha in response to 90 kg/ha N supplied in the form of urea and 60 kg/ha P_2O_5. This implies 10.6 kg of rice paddy per kilogram of nutrients (N + P_2O_5) applied. Second, rice yields increased even further when part or all of the N was supplied in the form of AS; the increases ranged from 1.2 mt/ha to 1.8 mt/ha. The average response to S applications ranged between 12 kg of rice paddy per kilogram of S when all urea was replaced with AS as a source of N and 54 kg of rice paddy per kilogram of S when only one-third of the total N (30 kg out of total 90 kg) was supplied by AS and the rest by urea.

Table 3.5. Average response of rice to urea and ammonium sulfate in South Sulawesi, Indonesia[a].

Nutrient source	Treatment N:P_2O_5:K$_2$O:S (kg/ha)	Average yield[b] (mt/ha)	Incremental yield over urea (mt/ha)	kg of rice per kg of applied S	% increase in average yield over urea
Urea	90:60:0: 0	3.22	–	–	–
AS	90:60:0:103	4.43	1.21	12	38
1/2 AS + 1/2 urea	90:60:0: 51	4.77	1.55	30	48
1/3 AS + 2/3 urea	90:60:0: 34	5.05	1.83	54	57

a. Derived from data reported in Ismunadji and Zulkarnaini (1978); originally from Mamaril et al. (1976).
b. Average of three locations including Kirukiru, Thung, and Lupakasi in Barru, South Sulawesi; using C_4-63 rice variety.

10. The model and the underlying assumptions for these estimates are described in detail in Chapter 7 which deals with estimating fertilizer S requirements in selected tropical countries of Asia, Africa, and Latin America.

There are important economic implications with respect to the social cost of lack of S use (Figure 3.6). In this case, the farmer lost a maximum of 1.8 mt/ha of rice paddy because the soil was deficient in S; the deficiency, however, could have been corrected with very little additional cost. The response to S may not be this high under farmers' field conditions on all the soils deficient in S, for all the crops and under all the agroclimatic systems. However, for a country the aggregate impact of inadequate S use over the extended period could be substantial in terms of lost crop production.

A food-deficit country like Indonesia cannot afford to forego such potentialrice production. The farmer experiences the immediate direct impact of lost output in terms of lost potential income. Indirectly, all the rice consumers suffer in terms of higher rice prices. However, if the government decides to keep rice prices lower than those determined by market forces, either the government (and hence taxpayers) must bear the cost in terms of price-subsidy to the consumer or price-support to the farmer, or the farmer must bear the cost in terms of lower income and reduced production incentives.

Economics of Fertilizer Sulfur

So far, almost no analytical or empirical research has been done on the economics of fertilizer S use in tropical countries. Some of the reasons for this lack of research are the following. First, the primary nutrients (N, P_2O_5,

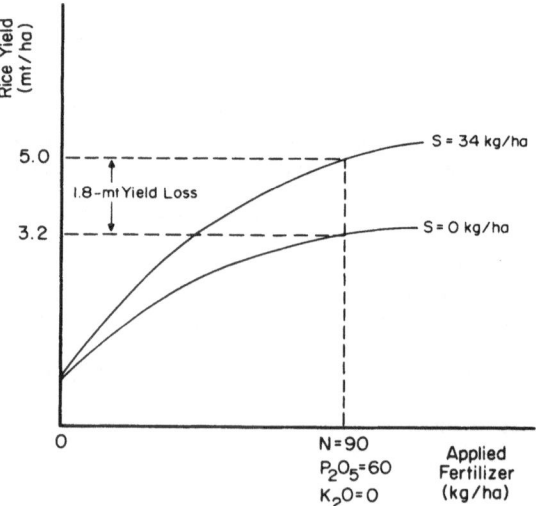

Figure 3.6. Loss in Paddy Rice Production Due to Sulfur Deficiency: Based on Experimental Results From Indonesia.

and K$_2$O) have always been given priority over S in fertilizer trials or demonstrations. Second, most people in the agricultural sector have not been aware of S deficiencies or have confused the symptoms with N deficiency symptoms. Third, the experimental data on crop response to S (under field conditions) have been inadequate for an appropriate economic analysis. Fourth, until very recently S was rather inexpensive, especially in developed countries. The rice response to S use, as reported earlier in Table 3.5, can be so large that the implied average value:cost ratio (after taking all cost components into account) is much more than 3, which FAO considers to be an adequate incentive for farmers to use fertilizers (Mudahar, 1978).

The economics of S use is not as straightforward as that of other nutrients. As with N, the crop uses most of the applied S in the first year. However, part of applied S, like N but unlike P$_2$O$_5$, is lost through leaching and/or volatilization, and part of it gets fixed in the soil. In this respect, S has some residual effects. However, very few empirical data are available on the fate of applied S in different tropical crops as to the proportions used by plants, fixed in the soil, or lost. Furthermore, there exists almost a fixed ratio between N and S (approximately 15:1 in plant protein) used by the plant. The plant cannot make productive use (in terms of quantity or quality of produce) of excess N or S unless the deficient nutrient is supplied exogenously.

As far as the economics of fertilizer S is concerned, there is a need for the following: (1) analysis of the economic impact of fertilizer S on economic development; (2) determination of economic returns to S use under different agroclimatic conditions; (3) comparative economic evaluation of existing (e.g., gypsum, elemental sulfur, SSP, and AS) and modified S-containing fertilizers; (4) economic analysis that accounts for the residual effects of S; (5) determination of delivered price of S to farmers by taking into account all cost components, including production, handling, storage, transportation, and other marketing costs; (6) economic evaluation of phosphogypsum as a source of S comparedwith natural gypsum and other S sources; (7) consideration of appropriate pricing of S content in S-containing fertilizers and its impact on the production, distribution, and use of these fertilizers; (8) evaluation of the economics of price and transportation subsidy on S sources; (9) analysis of the economics of indigenous S sources as fertilizers as opposed to imported S or S-containing fertilizers; and (10) economic assessment of S resources in developing tropical countries.

4 Sulfur in Plant, Animal, and Human Nutrition

The primary purpose of this chapter is to evaluate the role of S in plant nutrition, including its functions, uptake, and impact on crop quality. In addition, the chapter also deals with the interaction of S with other plant nutrients, recovery of applied fertilizer S, and the importance of S in animal and human nutrition.

Function of Sulfur in Plant Nutrition

Although the S content of plants is generally similar in order to that of P content, application of S has not received as much attention as has P application. This is due to incidental additions of S from many sources and the fact that S deficiency is masked by deficiencies of N and P. In the tropics S has received less attention than P because of the greater ease of mobility and availability of sulfate than of phosphate. Because of the P fixation and universal P deficiency in tropical soils, fertilization with phosphates, primarily SSP, became an accepted practice and S deficiency was overlooked even where it was apparent. Sulfur is required for the following functions in plants (Beaton and Fox, 1971; Blair, 1979):
1. Synthesis of three essential S-containing amino acids − cysteine, cystine, and methionine − which are essential components of proteins.
2. Formation of chlorophyll.
3. Activation of certain proteolytic enzymes such as papainases.
4. Synthesis of certain vitamins (biotin, thiamine, and vitamin B_1), glutathione, and coenzyme A.
5. Formation of glucosides which are essential components of oils found in onions, garlic, and cruciferous plants (mustard, for example).
6. Formation of certain disulfide linkages such as sulfhydryl (SH group), which besides giving pungency to oils also imparts resistance to drought and cold.
7. Formation of ferredoxin, an iron-containing plant protein that acts as an electron carrier in the photosynthetic process and is also involved in N fixation by both nodule bacteria and free-living bacteria.
8. Activity of ATP sulfurylase − an enzyme that functions in the metabolism of S.

Uptake and Translocation of Sulfur

Plants generally absorb S as the sulfate ion. In the pH range to which roots are normally exposed, S uptake is not very pH sensitive. Hendrix(1967) found the highest uptake rate by beans at pH 6.5. Most minerals scarcely affect the absorption of sulfate-S by plant cells. Selenate, however, depresses the absorption of sulfate-S because of the close relationship of the ions (Leggett and Epstein, 1956). Probably both the ions compete for the same site on the carrier. The actual uptake mechanism is not well understood, though there is a good deal of evidence that there is only one mechanism and its selectivity depends on the concentration of sulfate ions in the nutrient solution.

Ansari and Bowling (1972), working with sunflowers, observed that sulfate is absorbed and translocated against an electrochemical gradient which suggests that sulfate uptake is an active process. Sulfate is translocated in an upward direction, but the plants have little potential for moving S upward (low mobility). Thus in cases of an extreme deficiency of S or low supply of S from the soil, the lower leaves may show good supply of S and appear green while the upper leaves will be chlorotic and deficient in S. This behavior, which is contrary to that of N, can be useful in distinguishing between N and S deficiencies.

There is considerable evidence that plants can use SO_2 from the atmosphere to meet their S requirements to a certain extent. Once SO_2 is absorbed through the stomata it is distributed throughout the plant. Noggle (1980) observed that the yield of crops was higher near the coal-fired thermal power plant of the Tennessee Valley Authority than away from the plant and so was the S content of the crops. He attributed nearly 40% of the S in plants to the fact that the leaves directly absorbed SO_2 from the atmosphere in the vicinity of the thermal plant. These results show that reduction in the SO_2 content of the atmosphere can adversely affect crop production.

The total S content in plant tissues varies among plant species (Table 4.1). Raising the sulfate ion content in the nutrient medium raises the S supply and increases the organic S content without raising sulfate content in the tissue. As soon as the S demand of the plant is met, the additional S is deposited as sulfate, which keeps the organic S level constant. However, in plant species capable of synthesizing mustard oils, organic S rather than sulfate-S is stored, which explains the relationship between S supply and mustard oil content. With the exception of plant proteins containing S glycosides, the bulk of organic S is in the form of cysteinyl and methionyl residues. Since these proteins have a definite composition, the N:S ratio varies generally within a narrow range from 30:1 to 40:1. However, chloroplast proteins and nucleic acid proteins have a narrower ratio of about 15:1, and they contain a higher content of S.

The N:S ratios have been used as a diagnostic tool for S deficiency. Dijkshoorn and van Wijk (1967) proposed a ratio of 17:1 for legumes and 14:1 for grasses. Pumphrey and Moore (1965) estalished a critical ratio of 11:1 for alfalfa tops. McNaught and During (1970) consider a ratio of 16:1 for

Table 4.1. Sulfur content of various plant species.

Plant species	S content in dry matter (%)
Cereals	
Wheat	0.17
Maize	0.17
Barley	0.18
Oats	0.18
Legumes	
Broad beans	0.24
Soybeans	0.32
Bush beans	0.24
Peas	0.27
Crucifers	
Rape	1.00
White mustard	1.40
Oil radish	1.70
Black mustard	1.00

Source: Deloch, as reported in Mengel and Kirkby (1982).

rye grass and 18:1 for white clover to be critical. Ratio of protein N to non-protein N is also recommended for detecting S deficiency in plants because S-deficient plants will have more nonprotein N. Some crops need more S than others, depending upon the nature of the crop, the variety, and the available amount of S in the soil.

Sulfur Uptake by Different Crops

The total S requirement of different crops depends on the plant species and the yield levels or total dry matter produced. Crops with a high production of dry matter, such as sugarcane and maize, have a high demand for S. A high S requirement is also characteristic of protein-rich crops (legumes, lucerne, and clover), crucifers and brassica. The S requirement of rapeseed is nearly three times that of cereals. Generally the S content of most of the plants is between 0.1%-0.3%. However, as high as 2% S in leaves has also been reported, whereas roots invariably have lower amounts of S. The highest amount of S is in leaves and the smallest in roots.

Table 4.2 and Appendix I give the S requirements of important crops of the tropical regions. Generally, crucifers (such as cabbage, radish, turnip, mustard, rape), legumes (such as lucerne, soybeans, groundnuts), onion, garlic, cotton, sugarcane, maize, millet, oil palm, coffee, and tea require high amounts of S, whereas the cereals not mentioned above require rela-

Table 4.2. Estimated nutrients (S and N, P, K) removed by different crops.

Crop	Crop yield (mt/ha)	Total nutrient removal (kg/ha)				References
		S	N	P	K	
Cereals						
Rice (IR-8)	9.1	15	–	–	–	Wang (1978)
Rice (Peta)	6.1	17	–	–	–	Wang (1978)
Rice	3.0	9	84	14	89	Malavolta (1979)
Rice	7.8	14	125	30	137	Malavolta (1979)
Wheat	5.4	23	153	26	150	Malavolta (1979)
Wheat	3.0	14	125	22	92	Malavolta (1979)
Wheat	8.0	20	–	–	–	Mitchel & Blue (1981)
Maize	4.5	26	–	–	–	Malavolta (1979)
Maize	10.0	21	–	–	–	Mitchel & Blue (1981)
Maize	5.0	19	170	35	175	Malavolta (1979)
Maize	12.5	37	298	55	247	Malavolta (1979)
Millet	2.7	20	–	–	–	Fritz (1972)
Sorghum	2.5	11	–	–	–	Fritz (1972)
Sorghum	2.5	7	65	10	48	Malavolta (1979)
Sorghum	8.9	43	280	44	186	Malavolta (1979)
Fiber crop						
Cotton (lint)	1.7	34	201	71	141	Malavolta (1979)
Cotton	4.0	28	–	–	–	Mitchel & Blue (1981)
Oilseeds						
Groundnuts	5.0	10	–	–	–	Fritz (1972)
Groundnuts	4.0	21	–	–	–	Mitchel & Blue (1981)
Groundnuts	3.0	24	323	31	170	Malavolta (1952)
Soybeans	3.0	21	–	–	–	Mitchel & Blue (1981)
Soybeans	3.3	25	–	–	–	Potash Institute of North America
Soybeans	3.0	23	300	40	115	Malavolta (1979)
Soybeans	4.0	28	363	31	132	Malavolta (1979)
Oil palm	18.0	20	–	–	–	Fritz (1972)
Oil palm	24.6	–	193	36	249	Malavolta (1979)
Sunflowers	2.2	10	–	–	–	Western Canada Fertiliser Association (1978)
Sunflowers	3.9	18	197	34	120	Malavolta (1979)
Coconuts	–	–	74	16	113	Malavolta (1979)
Rapeseed	2.0	23	118	23	77	Western Canada Fertiliser Association (1978)
Sugar crops						
Sugarcane	100.0	22	–	–	–	Fritz (1972)
Sugarcane	224.0	96	403	76	567	Malavolta (1979)
Fruits						
Pineapples	65.0	11	–	–	–	Fritz (1972)
Bananas	35.0	5	–	–	–	Fritz (1972)
Bananas	30.0	–	627	69	1 390	Malavolta (1979)

Table 4.2. Continued.

Crop	Crop yield (mt/ha)	Total nutrient removal (kg/ha)				References
		S	N	P	K	
Stimulants						
Tobacco	2.0	12	–	–	–	Jordan & Reisenauer (1957)
Tobacco	3.0	20	–	–	–	Mitchel & Blue (1981)
Tobacco (flue-cured)	3.4	21	141	13	239	Potash Institute of North America
Tobacco (burley)	4.5	50	280	15	246	Potash Intitute of North America
Coffee	2.0	27	253	19	232	Malavolta (1979)
Cacao pods	9.0	6	–	–	–	Fritz (1972)
Tubers & Roots						
Cassava	19.0	8	113	11	79	Malavolta (1979)
Cassava	45.0	15	202	32	286	Malavolta (1979)
Potatoes	33.6	14	169	34	197	Western Canada Fertiliser Association (1978)
Potatoes	40.0	11	200	8	220	Malavolta (1979)
Potatoes	56.0	25	302	44	508	Malavolta (1979)
Beans						
Faba beans	3.4	12	204	21	81	Western Canada Fertiliser Association (1978)
Field beans	1.0	25	102	9	93	Malavolta (1979)
Peas	2.8	11	184	17	98	Malavolta (1979)
Vegetables						
Onions	34.0	25	–	–	–	Jordan & Reisenauer (1957)
Onions	37.0	34	133	22	177	Malavolta (1979)
Cabbages	84.0	64	280	31	249	Malavolta (1979)
Cabbages	34.0	45	–	–	–	Jordan & Reisenauer (1957)
Forages & Hays						
Alfalfa	22.4	57	672	58	558	Malavolta (1979)
Clover	13.4	34	336	44	335	Malavolta (1979)
Clover hay	6.0–9.0	17–22	–	–	–	Whitehead (1964)
Grass hay	6.0–9.0	9–13	–	–	–	Whitehead (1964)

tively smaller amounts of S. For normal yields, the crops with high S requirements need 20-45 kg S/ha; the crops with medium S requirements need 15-35 kg S/ha. On the basis of S research in Australia, Spencer (1975) has suggested the following S requirements in S-deficient areas for different crops:

Groundnuts	5-10 kg/ha
Cereal grains	5-20 kg/ha
Cotton	10-30 kg/ha
Sugarcane	20-40 kg/ha
Rape	20-60 kg/ha
Lucerne	30-70 kg/ha
Cruciferous forages	40-80 kg/ha

The uptake of S by crops depends not only on the S content of the plant but also on the expected yield level. Thus, doubling the crop yield may also double the S requirement of the crops. Likewise, increasing the intensity of cropping will create greater demand for S from the soil or from external sources such as fertilizers and manures. For example, the S content of rice may vary from 0.034% under an S deficiency to 0.16% under an S sufficiency or nonresponsive condition. The rice yields may also vary from 0.75 to 8.0 mt/ha and even more. This may result in S requirements that vary from 0.26 to 12.8 kg/ha, and even higher. Sulfur requirements by various crops under subsistence (low yield) and commercial (high yield) farming conditions are reported in Figure 4.1.

In field crops it is sometimes difficult to distinguish between S deficiency and N deficiency. In this instance a leaf analysis can be invaluable. In S-deficient plants the sulfate-S levels are very low, whereas amide-N and nitrate-N are increased. This contrasts with N deficiency where soluble N is depressed and the sulfate level is normal. In plants suffering from S deficiency, the rate of plant growth is reduced. Generally the growth of shoots is more affected than the root growth. Frequently the plants are rigid and brittle, and the stems remain thin. In contrast to N deficiency, chlorotic symptoms occur first in younger leaves, whereas the older leaves remain green. This indicates that the younger leaves depend on the S supplied by the root system directly. On the other hand, with N deficiency the N from older leaves is transferred to younger leaves, and the older leaves become yellow.

Plants are comparatively insensitive to high sulfate-S concentrations in the nutrient medium. When the sulfate concentrations are in the order of 50 ppm, as in saline soils, the plant growth is adversely affected. The critical concentration of SO_2 in the atmosphere, beyond which it may be toxic to plants, is in the range of 0.5-0.7 mg SO_2-S/m^3.

Sulfur Application and Crop Quality

Sulfur application increases the S concentration in plants. The S content of berseem (Trifolium alexandrinum), alfalfa, wheat, maize, groundnuts, soybeans, raya (Brassica juncea), and several oilseed crops has been reported

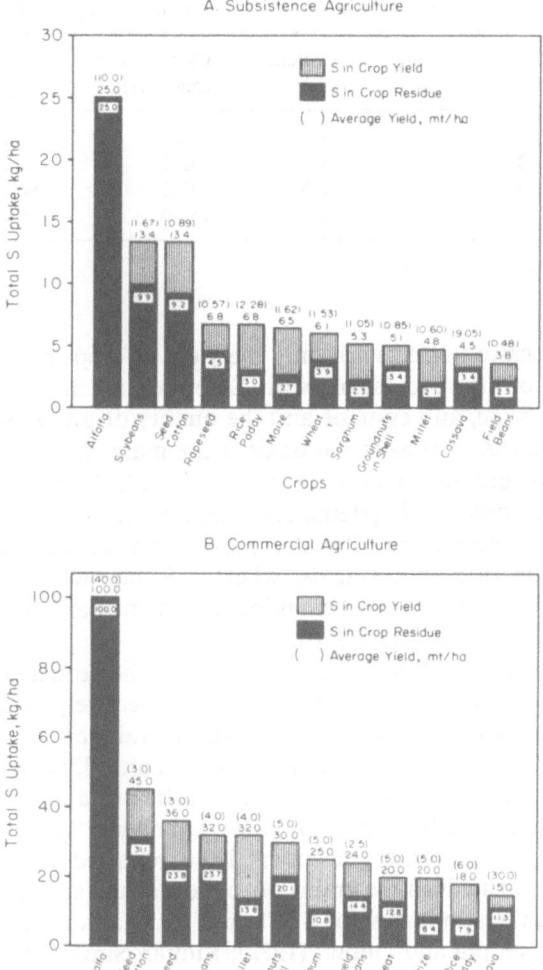

Figure 4.1. Effect of Cropping Pattern and Agricultural Modernization on Average Sulfur Uptake.

to increase with the application of S (Dev and Kumar, 1982). It was also found that in soybeans the distribution of fertilizer S was 29%, 33%, 17%, and 20% in grain, leaves, stem, and pods, respectively. Chahal and Virmani (1974) reported that when S was applied in the root zone of groundnut plants its relative distribution was 30.0%, 27.1%, 7.1%, and 35.8% in leaves, stem, pegs, and pods, respectively. The pattern of distribution was different when S was applied as foliar sprays.

Table 4.3. Effect of sulfur on yield and chemical composition of groundnuts.

S applied (ppm)	Weight of nuts (g/pot)	Protein (%)	Cysteic acid (mg/g N)	Methionine (mg/g N)	Oil (%)
0 – NPK	34.8	29.0	142.0	51.2	45.2
0 + NPK	43.9	29.7	145.2	52.2	46.2
50 + NPK	48.9	30.3	147.3	54.2	48.8
100 + NPK	52.8	30.6	151.0	55.0	49.6

Source: Chopra and Kanwar (1966).

Sulfur application affects not only the yield of crops but their quality as well because of its association with S-containing amino acids such as methionine, cystine, and cysteine and the quality of proteins. Many studies have shown that lack of S-amino acids is the main factor limiting the biological value of proteins. Chopra and Kanwar (1966) reported a significant increase in the content of cysteine and methionine by the application of S to groundnuts. Application of N influenced the protein content but not the content of S-bearing amino acids, whereas S improved both (Table 4.3). Similar beneficial effects on groundnuts and mustard were reported by Singh, Subbiah, and Gupta (1970).

Application of 20 ppm S as gypsum increased the protein and methionine content of groundnuts by 8.4% and 21.0%, respectively, and 50 ppm S increased protein and methionine in mustard grain by 6.3% and 10.7%, respectively (Kanwar and Randhawa, 1974). Aulakh, Dev, and Arora (1976) also reported that application of S increased the protein and S content and decreased nonprotein S in alfalfa; however, higher application of S increased both. Many Indian scientists have reported the increase in S-containing amino acids in the grains of soybeans, mustard, mung beans, and peas following the application of S [Chopra and Kanwar, (1966); Arora and Luthra, (1971); Gupta and Gupta, (1972); Kumar, Singh, and Singh (1981)].

Sulfur fertilization improved the oil content in mustard, groundnuts, and soybeans (Chopra and Kanwar, 1966; Dixit, as reported in Dev and Kumar, 1982). The increase in oil content of mustard was about 12% (Pasricha and Randhawa, 1973). Sulfur application also affects some other quality characteristics of crops. Ruhal, as repored in Dev and Kumar (1982), found that S decreased the watersoluble carbohydrates in groundnuts but increased them in wheat. Saroha and Singh (1979) observed a 5.6% increase in sugar content and a 5.8% increase in recovery of sugar from sugarcane.

On the basis of pot experiments for raya (Brassica juncea), Singh and Singh (1978) analyzed the role of S in the formation of glycosides, which on hydrolysis produce higher amounts of oil as well as allyl-isothiocynate, a compound that is responsible for pungency in oil. The source of applied S was potassium sulfate which was labeled with ^{35}S at a rate of 0.25 mCi/g of S. The allyl-isothiocynate value increased with increase in S dose up to 90 ppm. Sulfur applications of 60, 90, and 120 ppm significantly increased the allyl-isothiocynate value over the control, whereas lower doses were not significant.

It can be stated that, besides increasing crop yields, S fertilization has the following favorable effects on the growth of plants:

1. Improves protein, both in amount and quality, in pulses, cereals, tubers, and oilseeds, which are staple foods of people in tropical countries of Asia, Africa, and Latin America.
2. Increases protein content and decreases N:S ratio and nitrate levels in forages and thus improves their quality.
3. Improves quality of cereals for milling and baking.
4. Increases oil content of oilseeds and other oil-producing crops.
5. Improves quality, color, and uniformity of the vegetable crops.
6. Improves crop management through its favorable effect on drought tolerance, winter hardiness, control of diseases and pests, and decomposition of crop residue.

Sulfur Interactions With Other Nutrients

In order to develop a sound fertilizer use and management policy it is essential to know the interactions between S and other plant nutrients. Interactions of S with other plant nutrients such as N, P, K, Ca, magnesium (Mg), boron (B), iron (Fe), molybdenum (Mo), Zn, copper (Cu), manganese (Mn), and selenium (Se) are of great practical importance in designing fertilizer supply strategies and developing new fertilizer technology; hence, no agronomic or fertilizer management practice can ignore them.

S and N Interactions

Because of the central role of S and N in the synthesis of proteins, the supplies of S and N in plants are highly interrelated. It is for this reason that large doses of N create a severe deficiency of S and vice versa. O'Connor and Vartha (1969) observed that large doses of gypsum reduced the yield of hay when the N status of the soil was unsatisfactory. Likewise Eppendorfer (1971) observed that large doses of N created a deficiency of S. Aulakh, Pasricha, and Sahota (1980b)also observed a similar effect on mustard in India (Figure 4.2).

Application of S in the absence of N decreased the N concentration inmustard plants, but when N was added, the effect was synergistic (Dev and Kumar, 1982). Similar results were reported for amide-N and S in sunflowers (Sharma and Dev, 1980). Sen and Lahiri (1960) found that uptake of N was considerably reduced under S deficiency in sesame. The relationship of

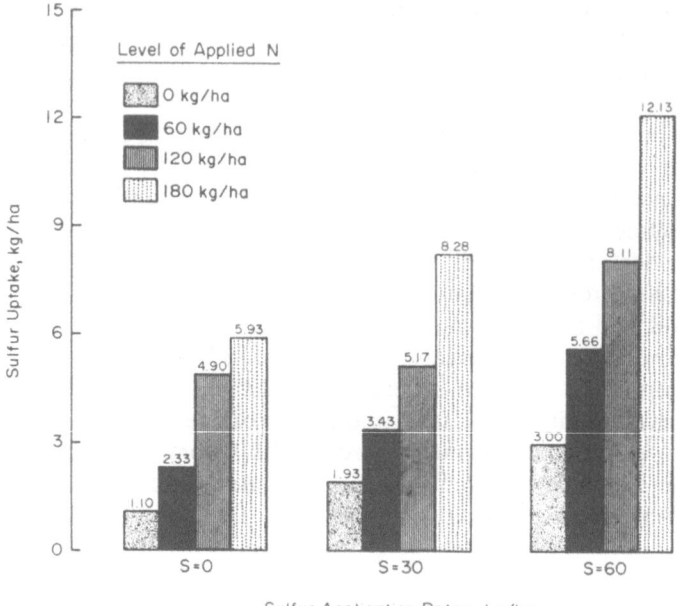

Figure 4.2. Levels of Sulfur Uptake by Indian Mustard at Different Levels of Nitrogen and Sulfur Applications (Supply Sources: Urea for N; Gypsum for S).

N:S ratios to plant health and growth has already been discussed. There is a very narrow range in the N:S ratio that ensures optimum yield and quality of the crop, and unbalanced fertilizer use adversely affects crop production.

S and P Interactions

Although S and P are evidently more loosely bound to each other in physiological terms than are S and N, the uptake of P has been stimulated in some cases by means of S fertilization. In addition, the P fixation in soils may be influenced by the acidity resulting from fertilization with S materials. The magnitude and direction of this influence, however, depend on the pH of the soil at the time of the S application.

On heavy fertilization with phosphate, the sulfate ions will be displaced from the adsorption sites and are apt to be lost in leaching. Liming soils of low pH will also have the tendency to increase their susceptibility to sulfate leaching. Thus in soils of low pH that have received lime and/or phosphate, use of NPK fertilizers containing sulfate may be a desirable way of reducing S deficiency.

Reports of Indian experience concerning the interactions of S and P on different crops are conflicting. Pathak and Bhardwaj (1968), Acharya and Subbiah (1971), Virmani and Gulati (1971), Venkateswarlu (1971), Rathee and Chahal (1977), Kumar and Singh (1980), and Marok and Dev (1980a and b) reported a positive interrelationship between the S and P contents and uptake in cotton, berseem, soybeans, rice, and wheat, whereas Aulakh and Pasricha (1977, 1979) observed that the simultaneous use of these two nutrients produced an antagonistic effect on chickpeas, lentils, and mung beans.

It may be observed that fertilization with P decreased the S content of the pulses but fertilization with S increased it. However, Aulakh, Pasricha, and Sahota (1980a) found a positive synergistic effect on P and S content of the foliage of groundnuts as well as on yield from a field experiment conducted on a soil deficient in both S and P. The source of S and P was SSP, which was compared with TSP and DAP, neither of which contained any S (Table 4.4). It may be observed that though the authors call it a synergistic effect actually the S content was lower in the presence of phosphate alone, indicating antagonism. The S content was increased by the SSP treatment because of the supply of S from SSP. Thus, the effect of P on S content is to depress it, which is expected. A similar antagonistic effect was reported by Aulakh, Pasricha, and Sahota (1977) in brown mustard and Indian mustard where the concentration of P decreased with the application of S. A negative interaction between S and P was also reported in berseem by Marok and Dev (1980b), who proposed a critical S:P ratio of 0.65 in berseem.

Thus it may be concluded that fertilization with phosphates generally decreases the uptake and concentration of S in the plant because of the antagonism. But this trend can be modified by the application of N, which stimulates the uptake of S.

Table 4.4. Effect of levels and sources of phosphorus on the yield and sulfur and phosphorus concentrations in groundnut foliage.

Source of P[a]	Pod yield (kg/ha)			Foliage dry matter (%)					
	20	40	Mean	Concentration of P			Concentration of S		
				20	40	Mean	20	40	Mean
Control	–	–	1 987	–	–	0.13	–	–	0.17
SSP	2 935	3 148	3 042	0.24	0.26	0.25	0.27	0.31	0.29
TSP	1 915	2 219	2 067	0.25	0.28	0.27	0.20	0.19	0.20
DAP	2 322	2 400	2 361	0.23	0.27	0.25	0.19	0.20	0.20
Mean	2 391	2 589	–	0.24	0.27	–	0.22	0.23	–

a. Levels of P_2O_5 were 20 and 40 kg/ha.
Source: Aulakh, Pasricha, and Sahota (1980a).

44

S and K Interactions

Appliction of S was found to increase the concentration of K in rice, mustard, and groundnut (Singh, 1971; Rathee and Chahal, 1977). A significant positive interaction between S and K was also observed in rapeseed by Aulakh and Pasricha (1978).

S and Ca, Mg Interactions

Pathak and Bhardwaj (1968) and Singh (1971) did not find any effect of application of S on Ca and Mg concentrations of berseem, rice, and alfalfa, but Aulakh and Pasricha (1978) observed a significant antagonistic interaction.

S and Micronutrient Interactions

Aulakh and Dev (1978) observed that S application increased Zn and Cu but had practically no effect on Fe and Mn content in lucerne. Kumar and Sing (1979) observed that in low doses S increased Zn uptake but in high doses caused antagonism in soybeans. According to Gupta and Mehta (1980), Fe concentration of berseem was increased by S fertilization.

S and Mo Interactions

Sulfur fertilization generally reduces uptake of Mo (Reisenauer, 1963a and b). Although phosphate is known to stimulate Mo uptake, SSP because of the presence of more sulfate than phosphate depresses Mo uptake (Gupta and Mehta, 1980). On the other hand, S deficiency may induce abnormally high toxic concentrations of Mo.

S and Se Interactions

The antagonism between S and Se is well known. In fact, as the Se content of the fertilizer increases, the S uptake and concentration in the plants decrease. Considerable evidence is available in India on these interactions, particularly with oilseed crops. Singh and Singh (1980) observed that the detrimental effect of 10 ppm Se in soil on yield and concentrations of S in rape and Indian mustard could be corrected by the application of 60 ppm S to soil. Because of the importance of both these elements in animal nutrition it is essential to know this relationship.

Uptake and Recovery of Applied Fertilizer Sulfur

Data on the recovery of S by the crops from fertilizer S applied to the soil are very scanty. Studies involving the use of ^{35}S in India and Africa reveal that the true recovery of S from S sources applied to the soil depends on a number of factors such as S status and nature of the soils; nature of the crop; management practices; and nature, dose, and method of application of the S-containing substance. These recovery values vary widely as is evident from some of the available information from selected countries in Asia, Africa, and Latin America.

Most of the results reported in Table 4.5 indicate that no more than 25% of the S recovered at low and medium application rates comes from the S source, and the rest comes from the soil. Even the residual effect of S is rather low since not more than 2%-3% fertilizer S is recovered by the second crop in rotation.

Sulfur in Animal Nutrition

The end products of animal production, including wool, meat, and milk, are protein-rich and thus require a high input of N and S in the diet. In this context S supply has important implications for animal nutrition and livestock production.

Adequate plant growth in forage crops requires that the N:S ratio be between 14:1 and 16:1. However, ruminants seem to perform satisfactorily if the N:S ratio is between 10:1 and 12:1. Thus, if the N:S ratio of the forages is to be made optimum for animal use, fertilization with S would be needed at a rate that might be above and beyond what is optimum for plant growth. High-yielding grass clover pastures need fertilization with phosphates, and if SSP is the source of P, the S needs are also met. However, the situation is different if TSP is used: an acute S deficiency develops, unless the soil, organic matter, or atmosphere supplies enough S, because the TSP supplies no S and the phosphates displace the sulfate.

There is voluminous literature on the responses of pastures and forage crops to S fertilization, which increases not only yield but also quality of the forage, and hence animal productivity (Tisdale, 1977; and Metson, 1973). The effect of S application has been studied more extensively in relation to forage quality and ruminant nutrition. Most of the research on S in Australia and New Zealand relates to forages and pastures for animal production. Tisdale (1977), while reviewing the work on forage quality and animal nutrition, summarized the information as follows:

1. Application of S increases the overall yield as well as the vitamin A content of alfalfa, the chlorophyll content of red clover, and the protein content of legumes and grasses. It decreases the N:S ratio of forages and the

Table 4.5. Fertilizer sulfur use efficiency.

Crop/soil	Source of fertilizer S	Rate of S applied to soil	S recovery from S fertilizer (%)	References
		Asia (India) (ppm)		
Groundnuts				
Samrala soil	Gypsum	10	12.1[a]	Subbiah & Singh
		20	13.1[a]	(1970)
Jaipur soil	Gypsum	10	10.3[a]	
		20	12.8[a]	
Mustard				
Ludhiana				
sandy loam	Gypsum	25	11.6[a]	Pasrichia &
		37.5	12.9[a]	Randhawa (1973)
Alfalfa				
Dark brown	Ammonium sulfate	20	12.7[a]	Shriniwas, Kataria
Red loam		20	11.3[a]	and Singh (1979)
Medium black		20	23.5[a]	
Maize				
Ludhiana	Ammonium sulfate	25	7.9[a]	Pasricha et al.
sandy loam		50	6.6[a]	(1977)
		Latin America (Brazil) (ppm)		
Rice	Ammonium sulfate	25	21.2[b]	Wang, Liem, and
		50	10.7[b]	Mikkelsen (1976a,
		100	5.6[b]	1976b)[c]
		Africa (kg/ha)		
Maize	Gypsum	20	2.8–14.0[a]	Bromfield,
	Sulfur	20	1.1– 8.2[a]	Hancock &
Beans	Gypsum	20	3.3–10.5[a]	Debenham (1982)[d]
	Sulfur	20	1.6– 6.0[a]	
Maize	Gypsum	20	11.0–40.5[b]	
	Sulfur	20	16.5–25.0[b]	
		U.S.A. (North Carolina) (kg/ha)		
Tobacco	Gypsum	4.48	17.0[b]	Kamprath, Nelson
		8.96	12.3[b]	and Fitts (1957)[e]
		17.92	7.2[b]	
		35.84	4.6[b]	
Cotton	Gypsum	4.48	73.5[b]	
		8.96	87.1[b]	
		17.92	47.0[b]	
		35.84	37.3[b]	

a. Isotopic recoveries calculated using radio isotopic techniques.
b. Apparent recoveries calculated by difference over the control.
c. Average of two rice varieties and N application rates.
d. Based on calculations of the authors' (published and unpublished) papers and reports obtained from ODA.
e. Apparent recoveries calculated from the data of the authors.

nonprotein nitrogen and nitrate content of grasses, and it generally improves the quality of alfalfa.

2. On soils that are low in S, the yield and quality of forages are improved through S fertilization.

3. Increasing the S content of forages in relation to protein N and reducing the N:S ratio to about 10:1 or 12:1 results in improvement of quality of feed, its use, and the performance of ruminants.

4. The total S levels in the ruminant diet should be between 0.18% and 0.25% for best animal performance.

5. Additional work by agronomists and animal nutritionists is needed to consider the merits of forage fertilization with S versus supplementation with S in animal feed.

Sulfur in Human Nutrition

Sulfur deficiency also has serious implications for human nutrition through its impact on crop yields and on quantity and quality of protein. Zake(1972) found that S fertilization increased the methionine content of finger millet to such an extent that the additional daily amino acid requirement of an adult was reduced from 1,325 to 725 mg/day.

Rice is the staple food in Asia, and any factor that affects the quality of rice creates concern for human nutrition. Ismunadji and Zulkarnaini(1978) reported results from experiments conducted in East Java, Indonesia, in1974 where soil was deficient in S. As reported in Table 4.6, application of S through AS slightly improved the crude protein and methionine content and hence the nutritional quality of rice.

In another study Ismuadji and Miyake (1978) observed that an S treatment increased the methionine content by 1.7-2.3 times that of the nonsulfur treatment. Wang (1978), from studies on rice in the swampy soils of the lower Amazon Basin, found that S deficiency caused not only drastic reductions in rice yield but also poor quality of grain, which was evidenced by reduction in head rice and an increase in chalky grain.

Table 4.6. Methionine and crude protein content of brown rice treated with urea or ammonium sulfate grown in Ngale, East Java, Indonesia, 1974 dry season.

Rice variety	Methionine (μM)		Crude protein (%)	
	Urea	AS	Urea	As
C$_4$-63	0.26	0.30	8.94	8.88
Pelita I/1	0.27	0.38	7.81	8.62
Pelita I/2	0.24	0.33	7.88	8.25

Source: Ismunadji and Zulkarnaini (1978).

In India, Das et al. (1975) observed that S application showed a favorable effect on the content of essential amino acids and S-containing amino acids in the grains of maize, wheat, and rice and thus maintained and improved the grain quality of these three important cereals (Table 4.7). Pasricha, Sharma, and Randhawa (1972) observed a beneficial effect of fertilization with S on the protein and oil content of groundnuts and mustard in Indian Punjab (Figure 4.3). Similarly, Singh, Subbiah, and Gupta (1970) reported a significant increase in oil, protein, and methionine content in groundnuts and mustard by the application of S.

The pulses, such as chickpeas (Cicer arietinum L.), mung beans (Phaseolus aureus), black gram (Phaseolus mungo L.), pigeon peas (Cajanus cajan), and lentils (Lens cultinaris L.), are important sources of proteins and S-containing essential amino acids in the Indian subcontinent and, to a certainextent, in Southeast Asia. Sulfur deficiency has been observed in these

Table 4.7. Effect of sulfur fertilization on protein: Total essential amino acids and sulfur-containing amino acid content of cereals in India[a].

Crop/ treatment[b]	Total essential amino acids (mg/100 g flour)	Sulfur-containing amino acids (mg/100 g flour)			Protein (%)
		Methionine	Cystine	Total	
Maize					
Site 1					
$N_{80} S_0$	3 678	249	169	418	8.75
$N_{80} S_{30}$	3 410	251	174	425	8.56
$N_{160} S_0$	4 357	250	200	450	10.50
$N_{160} S_{30}$	4 596	270	209	479	11.00
Wheat					
Site 1					
$N_{160} S_0$	6 406	229	280	509	17.27
$N_{160} S_{30}$	6 672	197	277	474	18.64
Site 2					
$N_{160} S_0$	5 975	221	203	424	15.16
$N_{160} S_{30}$	5 592	205	270	475	15.90
Rice					
Site 1					
$N_{160} S_0$	5 011	284	203	487	11.13
$N_{160} S_{30}$	5 021	306	229	535	12.14
Site 2					
$N_{160} S_0$	3 643	217	142	359	8.15
$N_{160} S_{30}$	4 412	213	225	438	11.31

a. Values are based on 100% recovery. Reproducibility of the analytical procedure is 3.5%.
b. Kilogram per hectare of N and S.
Source: Das et al. (1975).

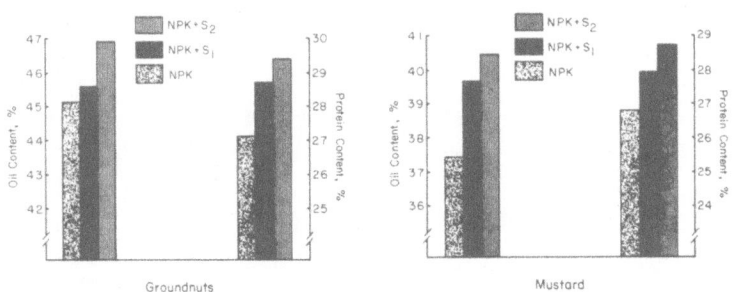

Figure 4.3. Effect of Sulfur on Protein and Oil Content of Groundnuts and Mustard in Ludhiana Sandy Loam Soil in Punjab, India.

crops in India. Dube and Misra (1970) reported a highly beneficial effect of S application on protein as well as on yield of some of these pulses and groundnuts. It has considerable practical implication for the vegetarian population of these countries. Pareek, Saroha, and Singh (1978) observed that an application of elemental S at the rate of 250 kg/ha increased the S uptake by black gram from 3.7 kg/ha for a control plot without S to 5.2 kg/ha on the treated plot in field experiments during 1972. This resulted in an increase in yield and improvement of quality of grain. Aulakh, Singh, and Arora (1977) reported that with potatoes an application of 25 kg of S, over and above the NPK (N : 120, P_2O_5 : 160, K_2O : 120 kg nutrient/ha), increased the yield of tubers by about 28%. Sulfur fertilization also influenced the quality of potato tubers and proteins in tubers, which increased from 56.7 kg/ha to 109 kg/ha with 50 kg S/ha. The corresponding levels of protein N in the tubers increased from 78.7 to 176.3 kg/ha.

These results clearly show that S favorably influences human nutrition in at least two ways: (1) through an increase in production of food and (2) through an improvement in quality of the food, particularly in the production of more protein and S-containing essential amino acids. Thus, any deficiency of S in tropical soils and crops will affect not only the food production but also its nutritive value; this has serious implications for human nutrition in these countries since vegetable sources account for a large share of daily caloric and protein intake.

5 Status, Diagnosis, and Determinants of Sulfur Deficiency

The purpose of this chapter is fourfold: (1) to evaluate the S status of soils in the tropical regions, (2) to discuss various sources and forms of S, (3) to analyze the appropriateness of various techniques for diagnosing S deficiency, and (4) to critically evaluate determinants of S deficiency in the tropics.

Sulfur Status of Soils in the Tropics

Delineation of Tropical Regions

The most common delineation of the tropics refers to the geographical area that extends from 23½° north of the equator to 23½° south of the equator. It includes the humid, subhumid, semiarid, and arid tropical regions. According to Dudal (1980), this represents 4.96 billion ha or 38% of the world's land mass, which is spread over Africa (43%); South America (28%); Asia (20%); Australasia (5%); and Central and North America (4%). Approximately 70.9% of the total land area in Africa (out of 3,011 million ha) and 70.4% in South America (out of 1,766 million ha) falls in the tropics.

Troll (1965) has used other criteria that consider (1) the mean monthly temperature of more than 18-233-C and (2) number of months with precipitation greater than potential evapotranspiration. According to this classification five tropical regions can be identified.

	Tropical Region	Months with P > PE
1.	Humid	9.0 – 12.0
2.	Subhumid	7.0 – 9.0
3.	Semiarid, wet-dry	4.5 – 7.0
4.	Semiarid, dry	2.0 – 4.5
5.	Arid	<2.0

P = precipitation; PE = potential evapotranspiration

The distribution of tropical regions in the world is shown in Figure 5.1. The most abundant group of soils of the humid tropics is the highly weathered and leached Oxisols and Ultisols. They constitute nearly 70% of the total area of humid tropics, while soils such as Mollisols, Vertisols, Alfisols, Andepts, Inceptisols, and Entisols cover the remaining 30% of the

51

52

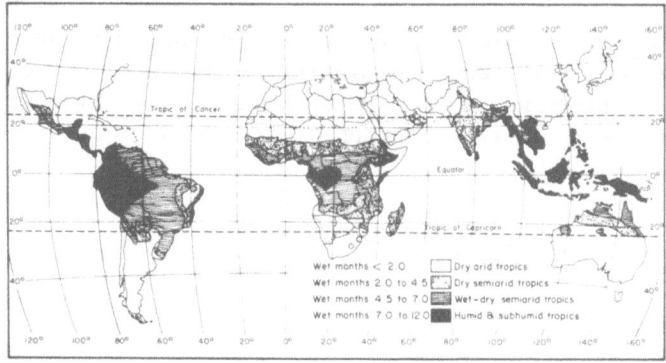

Figure 5.1. Geographical Distribution of the Humid, Subhumid, Arid and Semiarid Tropics in the World.

areas (Table 5.1). Swindale (1982), on the basis of FAO maps and Troll's classification, has estimated that total area under semiarid tropics with 2-7 wet months (precipitation > potential evapotranspiration) and 5-10 dry months equals about 1.8 billion ha (Table 5.2).

The semiarid tropical region, which includes nearly 49 countries of tropical Asia, Africa, and Latin America, covers much of the African continent, stretching in a broad band from west to east below the Sahara Desert and including much of eastern and subcentral Africa. In Asia the semiarid tropical region includes most of India, northeastern Burma, and Thailand; it also includes most of the northern quarters of Australia. Nearly all of Mexico, and large portions of Venezuela, Guyana, Surinam, French Guinea, Brazil, Paraguay, and Bolivia lie within this region. More than 700 million peo-

Table 5.1. Estimated area of major soil groups in the humid tropics.

Soil group/order	Climate (million ha)			
	Rainy	Seasonal	Wet-dry	Total
Mollisols and Vertisols	23	56	119	198
Alfisols, andepts, and inceptisols (moderately weathered)	4	34	90	128
Oxisols and ultisols (highly weathered and leached soils)	931	1 084	474	2 489
Shallow soils and dry sands and entisols	81	105	170	356
Alluvial soils (entisols)	146	124	71	341
Total	1 185	1 403	924	3 512

Source: President's Science Advisory Committee (1967).

Table 5.2. Estimated area of major soil groups in the semiarid tropics[a].

Soils (FAO/USDA soil taxonomy)[b]	Sandy	Total[c]	Sandy as % of total
	(million ha, approximate)		
Fluvisols (fluvents)	5.514	91.069	6.1
Arenosols (psamments)	201.731	211.195	95.5
Andosols (andepts)	5.034	10.783	46.7
Vertisols (vertisols)	2.533	132.436	1.9
Solonchaks[d]	0.455	5.754	7.9
Solonetz[e]	2.318	10.905	21.3
Kastanozems[f]	13.039	32.033	40.7
Phaeozems[g]	–	14.457	–
Cambisols (tropepts)	3.842	70.469	5.5
Luvisols (alfisols)	116.546	282.484	41.3
Planosols[h]	4.865	37.558	13.0
Acrisols (ultisols)	42.082	194.762	21.6
Nitosols[i]	11.487	107.476	10.7
Ferralsols (Oxisols)	43.122	606.018	7.1
Total	452.567	1 807.398	25.0

a. Derived from Swindale (1982), originally from FAO/UNESCO soil map of the world.
b. Names in parentheses are approximate equivalents according to Dudal (1980).
c. Includes sandy, silty, and clayey soils.
d. Solonchaks (salorthids and saline phases).
e. Solonetz (natriboralfs, natrabolls, etc.).
f. Kastanozems (haplustolls, arguistolls, calciustolls [except salorthids]).
g. Phaeozems (hapludolls).
h. Planosols (albaqualfs, albaquults).
i. Nitosols (paleudults, paleustults, paleudalfs, paleustalfs).

ple are estimated to live in the semiarid tropics, with 55% of them in India. This is a very important agricultural region, and it has been known to experience S deficiency. The cases of S deficiency reported from most of west Africa and the Indian subcontinent have occurred within this region.

Sulfur Status of Tropical Soils

The total S content of tropical soils is generally lower than that of temperate zone soils because of the lower organic matter content and greater leaching in the tropics. A summary of the data on total S content of soils from tropical areas is given in Table 5.3. Thus, while considering the problem of S in the tropical agriculture, it is essential to consider all the areas from arid to semiarid to humid tropics.

It is evident from Table 5.3 that there is only scanty information about the S status of soils of the tropical developing countries. Generalization is not advisable because of the very wide variation in soil S. Some of these variations may be due to differences in methods of analysis, but others are

54

Table 5.3. Total sulfur values for a range of soils from tropical regions (ppm in oven-dry soil)[a].

Region/location/soil group	Total S (ppm S)	Organic S (ppm S)	Sulfate-S (ppm S)	Adsorbed-S (ppm S)	Organic S as % of total-S	Reference
ASIA						
India						
Andhra Pradesh	112–275	–	–	–	–	Venkateswarlu, Subbiah, and Tamhane (1969)
Gujarat (Baroda & Kaira)	42–113	18–52	4 –22	–	–	Reddy and Mehta (1970a, c)
Uttar Pradesh (U.P.) (Soils of Tarai, Alluvial, Bundel Khand, and Vindyan regions of U.P.)	93–189	29–101	9 –52	–	–	Bhan and Tripathi (1973)
Subsoils of the above soils	102–169	15–78	5 –42	–	–	Bhan and Tripathi (1973)
Punjab, Haryana and Himachal Pradesh						
Soil group 1 (climatic zone 1)	193–308	106–244	1.4– 5.1	–	72.3	Kanwar and Mohan (1964)
Soil group 2 (climatic zone 2)	99–173	32– 60	3.1– 6.3	–	35.9	Kanwar and Mohan (1964)
Soil group 3 (climatic zone 3)	112–192	26– 50	1.8–18.1	–	22.0	Kanwar and Mohan (1964)
Soil group 4 (climatic zone 4)	128–168	22– 43	2.5–15.6	–	21.8	Kanwar and Mohan (1964)
Soil group 5 (climatic zone 5)	102–247	25– 34	2.5–41.4	–	18.8	Kanwar and Mohan (1964)
Mean						
a. Acid soils	242	175	3.4	–	72.3	Kanwar and Mohan (1964)
b. Alkaline soils	159	35	11.5	–	22.1	Kanwar and Mohan (1964)
c. Overall mean	183	76	9.1	–	41.7	Kanwar and Mohan (1964)
Rajasthan	750	53	76	–	7.0	Ruhal and Paliwal (1978)
Rajasthan (all soils)	91–386	60–298	22–83	–	22.2	Shukla and Gheyi (1971)
1. Serozen, 0–16 cm	271	90	20	–	20.9	Joshi, Choudhari, and Jain (1973)
2. Alluvial, 0–15 cm	300	160	40	–	53.8	Joshi, Choudhari, and Jain (1973)
3. Noncalcic brown, 0–20 cm	350	230	50	–	65.7	Joshi, Choudhari, and Jain (1973)
4. Desert, 0–15 cm	449	185	205	–	41.2	Joshi, Choudhari, and Jain (1973)

Table 5.3. Continued.

Region/location/soil group	Total S (ppm S)	Organic S (ppm S)	Sulfate-S (ppm S)	Adsorbed-S (ppm S)	Organic S as % of total-S	Reference
5. Brown, 0–20 cm	350	130	89	–	37.1	Joshi, Choudhari, and Jain (1973)
6. Red loam, 0–10 cm	375	150	55	–	50.6	Joshi, Choudhari, and Jain (1973)
7. Hilly, 0–20 cm	303	250	23	–	82.5	Joshi, Choudhari, and Jain (1973)
8. Greyish brown, 0–20 cm	250	140	15	–	56.0	Joshi, Choudhari, and Jain (1973)
9. Medium black, 0–15 cm	355	170	176	–	47.8	Joshi, Choudhari, and Jain (1973)
10. Yellow brown, 0–15 cm	375	320	12	–	67.3	Joshi, Choudhari, and Jain (1973)
AFRICA						
Malawi	35– 139 (66)	–	–	–	–	Laurence, Gibbons, and Young (1976)
Nigeria	38– 52 (43)	–	–	–	–	Bromfield (1972)
Cameroon	18– 132	–	–	–	–	Watson (1964)
Chad & Ivory Coast	200–300 (70)	–	–	–	–	Dabin (1972)
Zambia and Zimbabwe	60– 100	–	–	–	–	Grant et al. (1964)
Zimbabwe						
Fersiallitic and paraferrallitic soils						
Coarse textured						
0–30 cm	30– 60	–	1.7– 9.0	–	–	Rowell and Grant (1977)
30–60 cm	36– 56	–	3.2– 12.4	–	–	Rowell and Grant (1977)
Fine textured						
0–20 cm	116–144	–	4.3– 41.2	–	–	Rowell and Grant (1977)
40–60 cm	93– 157	–	2.7– 49.7	–	–	Rowell and Grant (1977)
Kenya						
Mount Kenya, humic nitosol						
Mean	370	359	11	2.0	97.0	Bromfield, Hancock, and Debenham (1982)
Range	(263– 540)	(238– 527)	(3 – 25)	(1.27 – 3.28)		

Table 5.3. Continued.

Region/location/soil group	Total S (ppm S)	Organic S (ppm S)	Sulfate-S (ppm S)	Adsorbed-S (ppm S)	Organic S as % of total-S	Reference
Kitale, rhodic ferralsol						
Mean	154	147	7	14.46	95.5	Bromfield, Hancock, and Debenham (1982)
Range	(105 – 187)	(102 – 182)	(1 – 17)	(1.04 – 1.54)		
Lake Victoria, orthic ferralsol						
Mean	144	132	12	0.94	91.7	Bromfield, Hancock, and Debenham (1982)
Range	(99 – 225)	(92 – 212)	(4 – 23)	(0.56 – 1.43)		
Coast, rhodic ferralsol						
Mean	85	83	2	0.80	97.6	Bromfield, Hancock, and Debenham (1982)
Range	(33 – 150)	(82 – 147)	(1 – 6)	(0.39 – 1.47)		
LATIN AMERICA						
West Indies						
Mollisol	360	–	–	–	–	Haque and Walmsley (1974)
Inceptisols	270	–	–	–	–	Haque and Walmsley (1974)
Regosols	210	–	–	–	–	Haque and Walmsley (1974)
Ultisols	120	–	–	–	–	Haque and Walmsley (1974)
Brazil (Virgin)	40 – 251 (103)	–	–	–	–	McClung, de Freitas, and Lott (1959)
Brazil (cropped)	27 – 67 (49)	–	–	–	–	McClung, de Freitas, and Lott (1959)
Brazil (cropped)	43 – 298 (166)	–	–	–	–	Neptune, Tabatabai, and Hanway (1975)
Colombia (Llanos)	394 – 405 (400)	–	–	–	–	Pedraza & Lora (1974)
Llanos	417	35	1.9	7.0	28.4	Guerarao & Orjuela (1979)
Savannas of Bogota	407	226	15.5	40.2	55.5	Guerarao & Orjuela (1979)

a. A dash (–) implies that the information was not available.

of pedogenic nature. For example, in Rajasthan arid zones, Joshi, Choudhari, and Jain (1973) and Ruhal and Paliwal (1978) reported high values for total S but lower values for the soluble sulfates and organic S. Probably most of the S appearing in the total S is insoluble sulfates or sulfates occluded in calcium carbonate. Interestingly, some of these soils are also reported to be responsive to the application of S.

Dabin (1972), in reviewing the S content of tropical African soils, concluded that the S content of ferralitic and tropical ferruginous desaturated soils of Ivory Coast, eutrophic brown soils of Cameroon, bottom soils of the savanna areas of Chad, bottom soils of the forest areas of Ivory Coast, and Vertisols or hydromorphic soils of Chad varied from 20 to 300 ppm with an average of 50-100 ppm, which is an indication of the low S content of these soils. The highest values were found in organic hydromorphic soils, whereas the others had low reserves. Mineral hydromorphic sandy soils and Vertisols are also low in S content. The sulfur:carbon (C) ratio is wider than 1:100, and the S:N ratio is wider than 1:10, which indicates that most of the S is in organic forms. Some of the typical values are shown in Table 5.4.

The Alfisols, which represent one of the most important soil groups of the west African region, are reported to be deficient in S. Kang et al. (1981) reported that S deficiencies were mor acute in the soils of the savanna zones than in those of the forest zones (Table 5.5). They attributed the greater S deficiency in the Guinea savanna zone of Nigeria to low S reserves, lower S retention, and sorption due to the sandy nature of the soils. Even the cropping pattern and the annual burning of brush and crop residue resulted in loss of soil S.

There are relatively few data about the soil S status of other African and Latin American countries (Do Nascimento and Morelli, 1980; Singh, Uriyo, and Kila sara, 1979). However, from these data it is evident that the soils of the tropics are not well supplied with S, and the deficiency of this nutrient can become a problem under intensive cropping. According to Sanchez and Cochrane(1980), approximately 52 million ha of high-base soils and

Table 5.4. Ratio of sulfur to other nutrients in West African soils.

Soil group	Country	S (ppm in soil)	Ratios		
			C:S	N:S	P:S
Ferralitic soils	Ivory Coast	0.38	130	10	1.0
Ferralitic soils	Central African Republic	0.26	150	10	3.9
Eutrophic brown soils	Cameroon	0.82	123	10	–
Bottom soils of forest areas	Ivory Coast	0.76	100	10	1.2
Bottom soils of savanna areas	Chad	2.40	175	10	4.5
Vertisols	Chad	0.54	148	10	3.5

Source: Dabin (1972).

58

Table 5.5. Sulfur content in some Nigerian soils (on dry-matter basis).

Soil attributes	Chemical composition (ppm)		
	Guinea savanna (low rainfall)	Derived savanna (medium rainfall)	Forest zone (high rainfall)
Total S	69.0	183.0	273.0
Total N	710.0	700.0	980.0
Organic P	86.0	117.0	185.0
Organic C	8 500.0	9 100.0	13 000.0
Heat-soluble S	2.6	4.3	7.0
$Ca(H_2PO_4)_2$-soluble S	2.8	3.5	5.9
Ratios of C:N:P:S	122:10:1.2:10	130:10:1.7:2.6	133:10:1.9:2.8

Source: Kang et al. (1981).

745 million ha of acid infertile soils of Latin America have an S deficiency problem.

Inorganic Sulfur

Sulfur is the 13th most abundant element in the earth's crust. Geochemically it is a constituent of many minerals of economic importance. It is present in most igneous rocks as sulfides (0.05%-0.3%). Its concentration in soils ranges from 0 to about 500 ppm (Ensminger, 1958; Starkey, 1950). As a result of oxidation during weathering, these primary forms of S are converted into sulfates, which may be precipitated as gypsum or many alkaline metal sulfates; or sulfates may be reduced to sulfides and elemental S under anaerobic conditions. Part of this may be carried to sea through drainage waters.

Sulfur in soil is present in both inorganic and organic forms, but the proportion of inorganic to organic S varies widely, depending on the nature of the soil, its depth, and the management system to which the soil is subjected. The common forms of inorganic S in the soil are (1) water-soluble sulfates of sodium (Na), K, Mg, Ca; (2) adsorbed sulfate on the surface of clay minerals and aluminum and iron oxides; (3) insoluble sulfates of Ca, barium (Ba), Fe, and aluminum (Al); and (4) sulfides or reduced forms of S.

Sulfate Forms and Sources

Although the level of soluble sulfate in the soils of humid regions is generally below 10 ppm, considerable fluctuations may occur. These variations are the result of mineralization from organic matter, leaching of soluble sulfates, uptake by plants, and sulfate addition from irrigation water and applied fertilizers.

In the well-aerated soils, inorganic S is present as sulfate. However, under anaerobic conditions, such as waterlogged marshy swamps, some of the inorganic S is present in reduced forms. Appreciable amounts of soluble sulfates are often found in subsoil horizons, and the occurrence of free gypsum in the deeper horizons of many semiarid soils is a well-known pedological phenomenon. Several forms of insoluble sulfates, such as Ca, Ba, and strontium (Sr), are associated with basic sulfates of Al, Fe, and calcium carbonate. Sulfate occurring in calcareous soils as a cocrystallized impurity in the calcium carbonate is probably the mot common in some soils of arid and semiarid regions and may account for 95% of the total S (Williams, 1975).

Some of the soil S comes from the primary minerals through weathering processes, and some comes from secondary sources by the accumulation of plant residues, roots, and other materials. Other important additions of S to soils come from the atmosphere, fertilizers, pesticides, and irrigation water.

Atmospheric Addition — The gaseous forms of S (for example, hydrogen sulfide [H_2S] and SO_2) generated from land life, ocean life, sea splash, and industrial activity become the atmospheric source of S, which enters the S cycle and contributes to S additions to soil and plants. The amount of S added from the atmosphere depends on the extent of industrialization, volcanic activity, and the generation of gaseous products that are brought down by rain to the land. Robinson and Robbins (1970) have calculated that about 66% of the total S generation on the earth is through oxidation of gaseous hydrogen sulfide produced by microbial decomposition of dead bodies; the remaining 34% comes from combustion of fuels and smelting plants and is the byproduct of the industry (Table 5.6). Sulfur in SO_2 form, generated naturally or artificially, enters the air, soil, rivers, and the ocean by rain and becomes a main source of S supply to the environment.

The atmosphere is a principal source of S for most soils. Annual sulfate-S additions in the precipitation of as much as 234 kg/ha and as little as 2 kg/ha have been reported (Reisenauer, 1975). In addition, considerable amounts of S may be absorbed by both plants and soils as SO_2 from air. The annual atmospheric contribution of S from natural generation to the land areas of the world has been estimated to be about 142 million mt; in contrast, the annual release from weathering of rocks is estimated as 2 million mt (Robinson and Robbins, 1970). The authors have also estimated that 73 million mt of S is released as SO_2 from artificial sources, 69% of this in the Northern Hemisphere and 31% in the Southern. These atmospheric additions could meet the S needs of the crops, but unfortunately they are not equally distributed; hence, S deficiencies in certain areas, particularly

Table 5.6. Natural and artificial sources of sulfur generation on the earth.

Sulfur source	Sulfur supply	
	Million mt	% share
Natural generation	142.0	66
1. H₂S from land life	68.0	32
2. H₂S from ocean life	30.0	14
3. Sea water splashes	44.0	20
Artificial generation	73.0	34
Coal combustion	51.0	24
Oil combustion	14.3	7
Smelting copper	6.4	3
Smelting lead	0.7	*
Smelting zinc	0.6	*
Total	215.0	100

* less than 1%.
Source: Derived from Robinson and Robbins (1970).

those that are landlocked and away from industrialized complexes, are more common.

The amount of SO_2 returned to the soil in the form of rain depends on the location of the industrial activities. Generally, it is many times higher within a 5-mile radius of the industry than away from it. Thus, the amount of atmospheric S added to the soil in the rural areas in the industrialized countries varies between 10 and 15 kg/ha/year. Even in the highly industrialized countries, however, the total amount of SO_2 emitted to atmosphere and returned to soil in rain will decrease because of regulations to control air pollution. Kiyoura (1982) has questioned the wisdom of applying very stringent pollution control measures, which reduce the SO_2 content of the atmosphere to such a low level as to reduce the crop yield and quality through S deficiency. The SO_2 content of the atmosphere in Japan decreased almost linearly from 0.06 ppm in 1967 to 0.015 ppm in 1978 in response to environmental regulations.

In the developing countries of the tropics, the atmospheric SO_2 is not a very important source of S because of the low level of industrialization. However, in some localized areas of developing countries, industrial activity may contribute to higher production of S, especially nearthe industry. In some areas where volcanic eruptions are common, SO_2 addition to soil can become significant. Recently, after the eruption of St. Helens in the state of Washington (U.S.A.), a significant increase in crop yields was observed. Some of this increase may possibly be attributed to the SO_2 emission from the volcano.

The S content of precipitation in the tropics, unlike that in the temperate zones, is usually very low. However, very little published information is available on this subject. Detailed studies by Bromfield (1974a,b,c) in northern Nigeria indicate that (1) the S content of precipitation at sites 350-880 km away from the sea ranged between 0.49 and 1.89 kg/ha/year and (2) the total annual accretion of S to soil is 2.35 kg/ha/year, which includes 0.65 kg S recovered as dust, 0.81 kg S as gaseous product, and 0.89 kg S in rain. The mean amount of S deposited in the rainy season was 1.14 kg/ha, which was hardly sufficient to replace the S removed by a typical yield of groundnuts obtained by farmers not using fertilizers. On the basis of this observation, Fox (1980) concluded that groundnuts production in the seasonally dry savannas of west Africa is being subjected to an S constraint.

Furthermore, Bromfield, Debenham, and Hancock (1980) observed that in central Kenya the amount of S added by rain ranged from 1.58 to 3.81 kg/ha (2.47 kg mean). On the basis of 26 years' rainfall data and a regression equation for the relationship between rainfall and S deposited, the mean amount of S deposited came to 1.71 kg/ha, and the range was 1.04-2.76 kg/ha. No such data are available from any other tropical developing country.

Fertilizers and Pesticides – The addition of S to soil through fertilizers is dependent on the intensity of fertilization, the nature of fertilizers, and management. The S content of important fertilizers is given in Table 8.1 and Appendix II. Generally with the introduction of high-analysis fertilizers, such as urea and TSP, the accretion of S is declining and S deficiency is increasing. Pesticides are also a source of S. However, with the introduction of more S-free pesticides, the amount of S added to the soil from this source is becoming less important. Soil amendments such as gypsum, pyrites, and elemental S, in use for reclamation of alkaline soils, are important sources of S.

Irrigation Water – Another important source of S addition is irrigation water, which acts as a source of sulfate supply as well as a means of leaching S from the soil. Thus, in irrigated areas the contribution of S through irrigation depends on the quality of water and management of irrigation.

Sulfates in Tropical Soils

In humid tropical regions, 70%-98% of the soil S is present in the organic form, which accumulates mostly in the surface horizons of the soil. The organic S must be converted to assimilable inorganic sulfate or sulfurous amino acids before it can be absorbed by plants. Sulfur mineralization is a microbial process that depends on such factors as moisture, aeration, temperature, soil acidity, organic S content, and N:S ratio of the added organic material. The S cycle resembles the N cycle in soils, and the mean C:N:S ratio in most soils is 125:10:1.2. However, there is a tendency for a wider ratio in the acid, lowbase soils, and a narrower ratio in the soils of the arid zones, calcareous soils and less weathered soils.

In the arid and semiarid regions, surface and subsurface accumulations of sulfates of Ca, Mg, Na, and K are common. Sulfate concentrations at some depth in the soil profile are also known to occur in nonarid regions where they are associated with horizons high in kaolinite, hydrated aluminum and iron oxides, or allophanes; their high capacity for retention of sulfate ions prevents the leaching of S from the soil. As much as 21 ppm of soluble S has been reported from the tropical soils (Fritz, 1972), and 280 ppm of S extractable with sodium acetate-acetic acid buffer has been reported from many soils of the southeastern United States (Jordan and Reisenauer, 1957). Sanchez (1976) reported that sulfateS in forms extractable with phosphate accumulated in Ultisols, Oxisols, and Andepts from Hawaii and ranged from 3.3 ppm in Ultisols to 134 ppm in Hydrandepts. The sulfate retention capacity of Costa Rican soils has been studied by Ramirez and Oelsligle (1978).

Under anaerobic conditions near the shore of brackish marine and fluviomarine sediments of total marsh areas, substantial amounts of S accumulate as sulfides or pyrites. When these waterlogged soils are drained, the sulfides, polysulfides, and elemental S are oxidized to sulfuric acid, which reduces the pH to a value as low as 2 or even lower, with all the consequent deleterious effects on plant growth.

Adsorbed Sulfates

The surfaces of iron and aluminum oxides and the weathered edges of clay particles contain ions that are not fully coordinated in the lattice. These ions complete their coordination shell with OH groups and water molecules. The presence of hydrogen ions favors the development of positive charges at the sites. These positive charges attract anions in the same way that surface negative charges attract cations.

When sulfate is added to soil, much of it is adsorbed. The common site for adsorption is on positively charged surfaces of iron and aluminum oxides and clay minerals. The sulfate adsorption capacity of soils has been shown to be affected by various factors as follows:
(1) Decreases with increasing pH (Ensminger, 1954; Kamprath, Nelson, and Fitts, 1956).
(2) Increases with increasing clay content (Neller, 1959; Chao, Harward, and Fang, 1962).
(3) Decreases with removal of iron and aluminum oxides (Chao, Harward, and Fang, 1962).
(4) Decreases with phosphate application (Metson and Blakemore, 1978).

(5) Decreases with organic matter content in the soil, hence, greater adsorption in subsoils of the tropics (Kamprath, Nelson, and Fitts, 1956; Chao, Harward, and Fang, 1962).

In arid soils, sulfate retention is commonly a question of gypsum solubility (Harward and Reisenauer, 1966). In acid soils, however, it is due to adsorption. The sulfate adsorption capacity of soils varies widely and is dependent on several soil properties. The tropical soils because of high iron and aluminum oxides or 1:1 clay minerals, or both, adsorb significant amouns of sulfate, which may be important from the point of view of storing S in the soil and thereby preventing its loss by leaching so that it becomes available for use by crops. Many of the low-pH soils of the tropics, particularly the coarse-textured soils, are low in available S supply and may need frequent applications of sulfate. But if they are underlain by a heavy-textured B horizon with higher adsorption capacity, soil S can become an important source of S.

Liming of acid soils will also increase the movement of sulfate out of the limed zone. Leaching by heavy rains may cause loss of S, but during the dry season and particularly in the tropics, the sulfate content of surface soil may increase because of mineralization of organic matter.

The type and amount of clay present in a soil determine the number of weathered-edge adsorption sites. Harward, Chao, and Fang (1962) found that the order of the amount of S retained by the clays was kaolinite > illite > montmorillonite. Aylmore, Karim, and Quirk (1967) have shown that sulfate adsorbed to iron and aluminum oxides is held more firmly than that held by kaolinite. This explains the higher degree of adsorption of sulfate in highly weathered soils of the tropics. Sulfate ions have a specific affinity for adsorption, and their presence almost completely prevents the adsorption of nonspecific nitrate and chloride.

The amount of adsorption of sulfate is dependent on the amount of the charge on the surface and hence on the pH value (Kamprath, Nelson, and Fitts, 1956). Furthermore, according to Couto, Lathwell, and Bouldin (1979), sulfate sorption by two Oxisols and an Alfisol of the tropics in Brazil was dependent on the pH of the equilibrium solution; the amount sorbed decreased as the pH increased with each soil.

Unlike adsorption of sulfate, the adsorption of undissociated ions such as the dihydrogen phosphate ion does not depend on the presence of a net positive charge; consequently, these ions are adsorbed to surfaces. In addition, the phosphate ions are very effective in displacing adsorbed sulfate and in reducing the capacity of the surface to adsorb additional sulfate. Aylmore, Karim, and Quirk (1967) found that solutions containing monopotassium phosphate desorbed some 20% more sulfate from clay and oxide surfaces than did water. Once the sulfate ions are desorbed and enter the solution, they are easily leached from the profiles. Thus, in the tropical soils with high phosphate fixation capacity, the application of phosphates results in greater mobility of sulfate and hence a greater tendency to loss by leaching with water. In soils with low adsorption capacity, added sulfate may not be retained and thus may move with the leachates (Metson and Blakemore, 1978).

The adsorbed sulfate is not free to move to plant roots by convective flow and diffusion. Therefore, in soils where there is not significant moisture movement throughout the profile, the availability of sulfate is lower in an adsorbing than in a nonadsorbing soil. Hence sulfate, which is available to plants, represents a balance between adsorption and leaching losses. Adsorbed S is generally higher in tropical soils than in temperate-zone soils. Phosphating and liming release sulfate from adsorbing sites and thus increase the possibility of its leaching as well as its availability to plants from the solution.

Organic Sulfur

The organic S is a part of organic matter and may be in the following forms: (1) ester sulfates; (2) bonded to C in a form other than amino acids; and (3) bonded to C as a constituent of amino acids.

Ester Sulfates

Ester sulfates are believed to be largely of organic sulfates containing ester linkages, such as choline sulfate, phenolic sulfates, and sulfated polysaccharides. These products of S are reducible by hydroiodic acid and alkali (Freney, Stevenson, and Beavers, 1972). Ester sulfates constituted 20%-65% of the total S in a group of six Brazilian soils and from 50%-62% in six soils from Iowa (Neptune, Tabatabai, and Hanway, 1975). In a wider range of soils in thetemperate regions, this fraction is found to vary between 30%-70% of total S (Williams, 1975). This fraction is generally unavailable to plants, but plantavailable S is released after breaking the linkage. Drying the soil breaks the linkage (Barrow, 1961) and can have dramatic effects on soil test results.

Carbon-Bonded Sulfur Compounds

Relatively little is known of the chemical nature of C-bonded organic S compounds in the soil (Freney, Melville, and Williams, 1970). However, the S-containing amino acids − cystine, cysteine, and methionine − have been isolated and may constitute up to 30% of the total organic S (Freney, Stevenson, and Beavers, 1972). The C:N:S ratio of soils approximates an average of 135:10:1.25 (Whitehead, 1964; Williams, 1967a, 1967b; Brook, 1979). A narrower C:N:S ratio is indicated in Indian soils because they contain very low levels of organic C. Moreover, the Walkley-Black method used in these studies for extracting the organic C underestimates C. Thus, it is evident that there can be considerable differences within each group. Dabin (1972) found a range of N:S ratios from 4:1 to 36:1 in soils of Africa. Neptune, Tabatabai, and Hanway (1975) found a variation from 3.4:1 to 12:1 in N:S ratios of Brazilian soils. There is a tendency for wider ratios in acid, low-base status soils and narrower ratios in calcareous and less weathered soils (Table 5.7).

Organic S is a reserve for plants, but it must undergo mineralization before becoming available to plants. Sulfur mineralization rates range from 1% to 10% per year (Sanchez, 1976). Bromfield et al. (1982) have estimated the rate of S mineralization to be 2.0%-2.3% annually at Zaria, northern Nigeria. Barrow (1961) found that immobilization occurs at C:S ratios greater than 200, and materials containing less than 0.15% S are immobilized. Andepts and other soils higher in allophane are also high in organic S, but plants growing on these soils are usually deficient in S because the association of organic matter with allophane results in low mineralization of S. Barrow (1961) also observed flushes of S, like flushes of N, resulting from mineralization when soils, previously dried, were wetted. However, the fate of the mineralized S may be different from that of nitrates because many soils have a greater capacity for adsorbing sulfate.

Like organic N, the amount of S in soils is also decreased by continuous

Table 5.7. Mean carbon:nitrogen:sulfur ratios in selected soils in the world.

Location	Soil class	C:N:S ratio	Reference
United States	Chernozem	114:10:1.6	Whitehead (1964)
	Black Prairie soils	119:10:1.6	Whitehead (1964)
	Podsolic soils	132:10:1.2	Whitehead (1964)
North Scotland	Noncalcareous	147:10:1.4	Whitehead (1964)
	Calcareous	113:10:1.3	Whitehead (1964)
Australia	Podsolic	155:10:1.4	Whitehead (1964)
	Acid soils	152:10:1.2	Whitehead (1964)
	Alkaline soils	140:10:1.5	Whitehead (1964)
West Indies	Regosols	156:10:1.2	Whitehead (1964)
	Mollisols	120:10:1.2	Haque and Walmsley (1974)
	Ultisols	117:10:1.1	Haque and Walmsley (1974)
	Inceptisols	110:10:1.0	Haque and Walmsley (1974)
	Mean of all West Indies soils	123:10:1.1	Haque and Walmsley (1974)
Nigeria	Mean of all non-fertilized soils	126:10:1.0	Sanchez (1976)
India	Alluvial soils U.P.	90:10:0.9	Bhan and Tripathi (1973)
	Alluvial soils Punjab & Haryana	89:10:1.2	Kanwar and Mohan (1964)
	Tea garden soils	99:10:1.2	Kanwar and Takkar (1964)

cultivation and cropping. McClung, de Freitas, and Lott (1959) observed that the organic S of virgin soils in Brazil decreased dramatically when the soils were cropped for 20-30 years (Figure 5.2). Similar results were also obtained by Bromfield (1972) in Nigeria. The results of McClung, de Freitas, and Lott (1959) also show that in tropical soils greater sulfate accumulates in the subsoils, where it can be stored and become available for deep-rooted crops.

Transformation of Organic Sulfur

Many factors affect the mineralization of organic S and transformations of S from one form to another. The organic S is mineralized through microbial processes, and the main factors that affect this transformation are as follows:

1. *Temperature* – According to Williams (1975) the optimum temperature for mineralization of S is 40-233-C. Oxidation of sulfur increases with a rise in temperature.
2. *Moisture* – Optimum mineralization occurs at 60% of the water-holding capacity (Chaudhry and Cornfield, 1967). Researchers from Australia have shown that when soils are dried before incubation the increase in mineralization of S is large, which may result in a flush of sulfate-S and could have an important implication for S fertilization. After dry periods for most of these soils, the S deficiency in plants disappears because of this flush of sulfate, and addition of S may not be necessary.
3. *S Content of Organic Matter* – Stewart, Porter, and Viets (1966) have shown that S mineralization occurs only when the S content of the straw is above 0.15%.
4. *Presence of Plants* – Mineralization is greater in the presence of plants than in the absence, probably because of the greater number of microorganisms in the rhizosphere.
5. *C:S Ratio* – Karwasra, as reported in Dev and Kumar (1982), observed

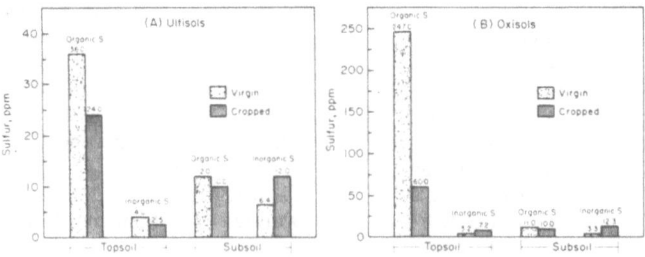

Figure 5.2. Effect of 20–30 Years of Cropping on Forms and Amounts of Sulfur in Two Soils of Brazil.

that mineralization of native S was greater in soils having wider C:S ratio. With the addition of sulfate, immobilization occurred. The addition of wheat straw led to immobilization of S, whereas berseem straw increased mineralization.

6. *Soil Properties* — The pH and calcium carbonate affect the mineralization of native and added S. In Sirsa soil (in the State of Haryana, India) a rise in pH from 7.4 to 8.2 increased organic S mineralization from 5.6% to 7.7%. Chopra and Kanwar (1968) observed that oxidation of added S was low in an acidic soil of Palampur (in the State of Himachal Pradesh, India) because of its lower pH. Addition of calcium carbonate enhanced the mineralization of added S. Under anaerobic and waterlogging conditions the S was immobilizedor transformed to sulfide and organic S (Dev and Kumar, 1982).

Diagnosis of Sulfur Deficiency

Plants suffering from S deficiency develop characteristic symptoms, mostly yellowing of young leaves, which are often confused with symptoms of N, Fe, and other nutrient deficiencies. Young leaves are light green to yellowish in color with lighter colored veins. However, in crops like tobacco, cotton, and citrus fruits the older leaves may be affected first instead of the young leaves. In sorghum and maize S deficiency is often confused with Fe and Zn deficiency. In legumes nodulation is reduced. Sulfur-deficient plants are generally small and spindling with short, slender stalks; they show poor growth and late maturity.

Soil tests and plant tissue tests are often used to determine the deficiency of S in soils and plants. Because of the difficulties encountered in determining S and the inadequate data on critical values of S for different crops, one of the tests alone can be used as an infallible guide. A combination of both soil and plant tissue tests, however, provides a good indication of the availability of soil S to the crop. There is still a need to correlate the soil and plant tissue tests with specific crops and soils. Determination of S in soils and plants is frequently considered a difficult task for three reasons:

1. Converting and isolating the total S or a specific fraction of S into forms suitable for analysis is often time consuming and leads to S losses.
2. Analysis for S after its conversion to sulfides or sulfates is generally laborious, particularly at low concentrations.
3. Instrumental methods of analysis for determination of S in grain and plant tissues are also not very satisfactory.

Several comprehensive reviews of methods for determining available S in soils have appeared in literature in recent years: Beaton, Burns, and Platou (1968), Reisenauer, Walsh, and Hoeft (1973), Tisdale and Nelson (1975), and Brook (1979).

Soil Testing Methods

The soil testing methods that use chemical extractants can be classified into the following three groups:
1. Those that extract readily soluble sulfate-S from the soil.
2. Those that extract the readily soluble sulfate and part of adsorbed sulfate-S.
3. Those that extract the readily soluble S, part of adsorbed sulfate, and part of organic S.

Readily Soluble Sulfate S − The common extractants are these:
1. Cold water.
2. Calcium chloride − neutral 0.15% $CaCl_2$ solution.
3. Lithium chloride − neutral 0.1 M LiCl solution.

In all the three methods the soil:extractant ratio is 1:5 and the extraction time is 30 minutes. These methods have generally given better results in soils of semiarid tropics and arid regions because of the higher amount of soluble sulfate in these soils.

Readily Soluble + Part of Adsorbed Sulfate S − The important methods are as follows:
1. Heat-soluble sulfur − sequential wet and dry heating of the soil (1:4 soilsolution ratio, pretreatment at 102-233-C for 1 hour using 100 ml water for extraction).
2. Monocalcium phosphate − 0.01 M $Ca(H_2PO_4)_2$ solution or 500 ppm P as $Ca(H_2PO_4)_2$ solution (1:5 ratio).
3. Potassium phosphate − 500 ppm P as KH_2PO_4 solution (1:10 ratio, pH 6.5).
4. Neutral ammonium acetate − 1 N NH_4OAc solution (1:2.5 ratio, 30 minute extraction).
5. Sodium acetate + acetic acid − NaOAc + HOAc solution (pH 4.8, 1:2.5 ratio, 30-minute extraction).

Readily Soluble + Part of Adsorbed Sulfate + Part of Organic S − The extractants normally used are as follows:
1. Sodium dihydrogen phosphate − 0.03 M $NaH_2PO_4 \cdot 2H_2O$ in 2 N acetic acid (1:5 soil-extractant ratio, 30-minute shaking).
2. Monocalcium phosphate − 0.01 M $Ca(H_2PO_4)_2$ in 2 N acetic acid (1:5 ratio, 30-minute shaking).
3. Sodium bicarbonate − 0.5 M $NaHCO_3$ (pH 8.5, 1:4 ratio, 1-hour shaking).

The summary of results available on critical values for available S by different methods for different crops is reported in Table 5.8. These data show great variability in critical values in different crops depending on the

Table 5.8. Critical levels of sulfur in soil for different crops and in different countries.

Extractant or methodology	Crop	Critical level of S in soil (ppm S)	Country	Reference
Water	Legume grass	3	Canada	Walker and Doornenbal (1972)
Ammonium acetate	Millet	6–7	Brazil	McClung, de Freitas, and Lott (1959)
Calcium chloride	Legume grass	3	Brazil	McClung, de Freitas, and Lott (1959)
	Sunflower	19	India	Marok and Dev (1979)
	Groundnut	5	India	Chopra and Kanwar (1966)
Potassium phosphate (500 ppm P)	Pastures	8	New Zealand	Cooper (1968)
	Pastures	4	Australia	Spencer and Barrow (1963)
	Maize	4	Nigeria	Kang and Osiname (1976)
Lithium chloride	Brassica	Variable	West Indies	Haque (1971)
Monocalcium phosphate (500 ppm P)	Alfalfa	10	U.S.A.	Fox et al. (1965)
	Maize	8	U.S.A.	Fox et al. (1965)
	Maize	4	Nigeria	Kang and Osiname (1976)
	Pastures	4–5	Australia	Andrew (1975)
	Cluster beans	10	India	Virmani (1971)
	Lucerne	9.3	India	Bansal, Sharma, and Singh (1979)
Sodium acetate and acetic acid	Alfalfa	12	U.S.A.	Harward, Chao, and Fang (1962)
	Maize	6	Nigeria	Enwezor (1976)
Ammonium acetate	Brassica	Variable	West Indies	Haque (1971)
	Maize	4	Nigeria	Kang and Osiname (1976)
	Peanut	6–7	Brazil	McClung, de Freitas, and Lott (1959)
Sodium bicarbonate	Cotton	10	U.S.A.	Kilmer and Nearpass (1960)
Monocalcium phosphate + 2N acetic acid	Alfalfa	10	U.S.A.	Hoeft, Walsh, and Keeney (1973)
Sodium dihydrogen phosphate + 2N acetic acid	Pastures	10	U.S.A.	Cooper (1968)
Hydrochloric acid (0.01 M)	Pastures	200	Australia	Andrew (1975)
A values	Cotton	15	U.S.A.	Nearpass, Fried, and Kilmer (1961)
A values	Alfalfa	10	U.S.A.	Harward, Chao, and Fang (1962)
A values	Sunflower	30	India	Marok and Dev (1979)

methods of extraction. This is not unexpected as the type of soil, method of sampling, extractant used, method and time of extraction, and chemical method of estimation of S all affect the result.

Dev and Kumar (1982), while reviewing S research in India, concluded that the suitability of a particular extractant depends on a number of factors such as soil type, pH, and the test crop. Chopra and Kanwar (1966) observed that heat-soluble S and 0.1 N calcium chloride-extractable S were highly correlated for groundnuts and berseem, respectively. The 0.15% calcium chloride, 500 ppm P as monocalium phosphate, magnesium acetate, and ammonium acetate and acetic acid were relatively more efficient than the other extractants for predicting S availability for berseem, cowpeas, and maize in India. Virmani (1971) found monocalcium phosphate a better extractant for predicting the S available to cluster beans, and the critical limit (i.e., the value above which no response to S additions was noted) was 10 ppm. Using the same method Bansal, Sharma, and Singh (1979) found that 9.3 ppm was the critical value for lucerne. McClung, de Freitas, and Lott (1959) reported that the critical level of sulfate-S in Brazilian soils from the Cerrado was 6-7 ppm extractable with ammonium acetate, and above this value no responses to S were detected. In Costa Rica soils, 8 ppm of sulfate-S was found to be critical for sorghum (Perez and Oelsligle, 1975).

Couto, Lathwell, and Bouldin (1979) reported that because of the greater adsorption of sulfate in soils of humid tropics, it is essential to consider the adsorbed sulfate in the subsoils for interpretation of responses to S. For tropical soils, extraction with monocalcium phosphate has become popular because this solvent easily extracts the adsorbed sulfate as well as organic S which determines the S reserves. Moreover, the extract obtained thereby is less colored, which facilitates determination of sulfate by turbidimetric methods. Tabatabai (1982) recommends use of 0.01 M monocalcium phosphate solution for extracting adsorbed sulfate in soil and use of the ion-chromatographic technique (Dionex model 10-ion-chromatograph) for estimation of S.

From the analysis of 30 rice soils, Islam and Ponnamperuma (1982) concluded that the critical concentrations of available S as determined by calcium phosphate, lithium chloride, ammonium acetate, and hydrochloric acid extraction methods were 9, 25, 30, and 5 ppm. Though the soil testing methods gave different values, all could be used to differentiate between the S-deficient and nondeficient soils.

Soil Tests and Sulfur Response Correlation

Relatively little work has been done on the correlation of soil tests and S responses in tropical countries, though considerable information exists on this subject for temperate zone soils (Probert and Jones, 1977). However, field and pot culture studies in Asia, Africa, and Latin America, as well as yield increases due to S treatment, indicate widespread S deficiencies in many of the developing tropical countries.

The studies relating to correlation of soil tests for available S with responses to S application can be grouped into two categories. The first is those that attempt to find a correlation between the degree of S deficiency and the soil test value for available S. They express the crop yields obtained without S as percentage of yield obtained with S. Instead of crop yield, S uptakes with and without S treatment are also compared. The second group consists of those that aim at establishing critical limits below which a response to S additions should be expected.

In the first case an attempt is made to determine the degree of S deficiency, whereas in the second case only the likelihood of such a deficiency occurring is determined. Most of the studies are greenhouse studies that, because of convenience and ease of manipulation, are in common use. No doubt they are valuable for some limited purposes such as determining relative efficiency of S-supplying substances, but they have serious limitations for extending the conclusions to field situations because of the following factors:

1. A small volume of soil is exploited by plants.
2. The amount of water used for watering the pots is many times more than that used in the field experiments. If distilled and deionized water is used, it may overemphasize the S deficiency, which normally would have disappeared because of inadvertent addition of sulfate from irrigation water. If tapwater or normal irrigation water is used, it may add more than normal amounts of sulfate, which may mask an inherent S deficiency. Yoshida and Chaudhry (1979) observed that when nonsulfur fertilizers and demineralized water were applied in pot culture trials on rice in the Philippines, an S deficiency was induced in potted plants on Lipa clay loam (an upland counterpart of Maahas clay) soil, which is not normally considered S deficient.
3. The temperature is generally higher in pot cultures than in the field; this may lead to greater mineralization of organic S reserves of the soil.
4. Greenhouse conditions prevent the accretion of atmospheric S, which would occur under normal field conditions.
5. Lack of conformity to actual soil profiles seriously limits the utility of pot culture tests since the combined effect of S in surface soil and S in subsoil may be different from the effect of S in either of them. In tropical soils, particularly those that have high amounts of adsorbed sulfates in the subsoil, pot culture trials with the surface soil may give an erroneous idea about the S status of the soil.
6. Most often in the case of pot culture, crops are not allowed to grow until maturity. As a result, the full impact of the S deficiency may not be realized. Blair (1979) observed that pot culture studies are of limited value because they are often short-term studies that fail to indicate the full ef-

fect of S deficiency on the crop. However, for understanding the relative but not absolute behavior of the soil and crop, pot culture experiments are important tools. Andrew, Crack, and Rayment (1974), while reviewing the experience with pot culture trials in Australia, concluded that use of pot culture results is limited in that the system may overemphasize the nutrient needs because of conditions under which plants are growing. However, these results are valuable when used in conjunction with the chemical analysis of the soil, full profile analysis, and consideration of environments.

Major problems concerning S studies are associated with analytical methods, sample collection and preparation, extraction analysis, and interpretation of results. Because of the rapid changes that S can undergo in the soil, it is essential that soil samples be collected immediately before the analysis is made, and long-term storage should be avoided. The role that S adsorbed in the subsoil may play in crop nutrition makes it necessary to rely more on field experiments for studying the need for fertilization with S in different crops and soils.

Biological Methods

In addition to chemical extractions a number of biological methods are also used to determine available S. According to Beaton, Burns, and Platou (1968) the biological methods are as follows:

1. Radioactive S is used to determine the amount of sulfate originally in the soil that is assimilated by the plant tissue. This is called the 'A' value.
2. The yield of nutrient curves is extrapolated to obtain the 'a' value, which is closely related to 'A' values.
3. Growth of algae is used to indicate the available S status.
4. Growth of Aspergillus niger is used as an indicator of inorganic sulfate.
5. Barley seedlings are used to extract available S (Neubauer method).
6. Soil is incubated to measure its capacity to convert organic S to inorganic sulfates.
7. Plant respiration is used to assess the degree of S deficiency by comparing the respiration curves with and without S.
8. Extractions of S from the root pads of turnips or wheat are used to estimate the short-term uptake of S.

Because these methods are time consuming and less practical, they are not commonly used. But there is much evidence that the methods also can serve as useful indices of available S. Mehlich (1970), using the Cuninghamella method, established S responses for tomatoes in Kenyan soils that were good indicators of field conditions. However, more critical studies of these methods are needed.

Tissue Testing for Sulfur Status

The sulfate-S content of plants has been used as a sensitive indicator of their S status. The various extractants include water, 2% acetic acid, trichloroacetic acid, hydrochloric acid, formic and hypophosphorus acids, sodium hydroxide, acetone, and ethanol. A detailed review of all of these is given by Beaton, Burns, and Platou (1968).

The analytical techniques include chromatographic, colorimetric, gravimetric, microbiological assay, radioactivity assay, spectrographic, titrimetric, and turbidimetric procedures. By the methylene blue method of Johnson and Nishita (1952) it is possible to determine sulfate-S directly in the dried plant-tissue samples without ashing.

The sulfate-S levels of 0.02%-0.03% in alfalfa, 0.02% in coffee, 0.013% in cotton leaves, and 0.02% in rapeseed leaves have been reported to be critical levels (Beaton, Burns, and Platou 1968). Total S concentration of leaves also has been used as an index of sufficiency or insufficiency of S in the plant tissue. Kamprath and Jones (In press) have presented considerable data from the United States indicating that when S content in the tissues was less than 0.15% in maize, 0.32% in soybeans, 0.20% in cotton, 0.10% in sugarcane, and 0.14% in wheat leaves, significant responses to fertilization with S occurred. However, the authors did not call these critical values.

While reviewing the efficiency of different methods for predicting S deficiency by plant and soil analysis, Murphy and Brogan (1981) concluded that, in over 100 field experiments conducted during the past few years in Ireland, plant analysis had not been successful in predicting S deficiency. However, the sulfate content of herbage at midseason did give a good indication of the S status of the herbage and could be used to predict late season S deficiency. Blair (1979) concluded that plant analysis for S appears to offer some promise. However, problems with the analytical procedures must be overcome before useful data can be obtained. The specific issues that need attention include the following:

1. *Sample Preparation and Analysis* – Establishing the critical levels of S without specifying the method of digestion and determination is of little practical value for setting levels of adequacy. The evidence for this conclusion comes from the data of Kang and Osiname (1976) who established 0.14% S as a critical level of S in the earleaf of maize in Nigeria. Daigger and Fox (1971) observed this value to be 0.24% S. The differences in the values obtained by the two groups of scientists could be explained by the ifferences in the methods of analysis. While Kang and Osiname (1976) used the sodium carbonate fusion method, Daigger and Fox (1971) used a mixture of nitric and perchloric acids. The following data from Sansum and Robinson (1974) illustrate that, depending on the method of digestion, different results are obtained with respect to S concentration.

74

Method	% S
Dry ashing	0.150
Wet ashing with nitric-perchloric acid treatment	0.207
Nitric-perchloric acid digestion for 1 hour after clearing	0.355

Often critical levels are established on shaky evidence. For example, De Freitas, Gomes, and Lott (1972) concluded that 200 ppm sulfate-S is the critical limit for coffee leaves. However, as pointed out by Blair (1979), the spread of the data indicates that this criterion is too insensitive to be a useful diagnostic tool.

2. *Sulfate Versus Total S in Tissue* — Lott, McClung, and Medcalf (1960) considered that sulfate-S is a better index of S deficiency in coffee than is total S. However, according to the observation of Anderson and Spencer (1950), sulfate may accumulate when there is a molybdenum deficiency, even though the plant may suffer from S deficiency. Thus, any criteria based on sulfate-S in plant tissue without consideration of the other nutrients could lead to erroneous conclusions. Freney, Randall, and Spencer (1982) were of the opinion that a better index of the S status of the plant is the proportion of sulfate-S to total S in the plant tissue. Very strong correlations exist between this index and yield, and the relationship is independent of plant age and N level. Grain analysis can also be used for diagnosis of S deficiency. A staining technique has been developed that can distinguish between low-S and high-S grain. Grain analysis for S is of great practical use because of the effect of S on the baking quality of wheat flour.

3. *Differences in Cultivars* — It is well established that different cultivars of the same species do vary in their S contents. For example, Fox, Kang, and Nangju (1977) observed that cowpeas cultivar Sitao Pole showed a critical value of 0.032% S and cultivar TV476-2E, 0.064% S. Thus the cultivar variation makes one value a useless criterion for the other.

4. *N:S Ratios* — Researchers have used N:S ratios also as a diagnostic tool. The ratio varies from 14:1 for graminaceous to 17:1 for leguminous specie sand 15:1 for most of the crops. It is believed that variations in the amount of N and S fertilizers used do not appreciably change the N:S ratios of the protein (Dijkshoorn and van Wijk, 1967). Dev and Saggar (1974), with 12 cultivars of soybeans in India, observed that S application at varying levels lowered the total N:S ratio and widened the protein N:protein S ratio. They also observed that these ratios varied from 12 to 16 in different varieties, which indicates great varietal differences. Bansal and Singh (1979) used N:S ratio for diagnosing S status of alfalfa.

Reneau (1981), on the basis of studies on S requirements of maize in Virginia (U.S.A.), concluded that the critical concentration of total S would be

0.18%, and the N:S ratio in the leaf opposite and below the ear leaf at silking would be 15. Islam and Ponnamperuma (1982) concluded that the critical N:S ratio was 15 in the shoot of rice at maximum tillering, 14 in the straw at maturiy, and 26 in the grain. The sulfate-S expressed as percentage of total S was 15% in the shoot at tillering and in the straw at maturity. The effects of varietal variations and stage of development or maturity of the leaf on S content and also N:S ratios make these criteria not very satisfactory for predicting an S deficiency.

From these discussions it can be concluded that both soil testing methods and plant tissue testing methods have limitations. However, the soil testing method may have certain advantages for tropical soils. In fact, the use of extractants like monocalcium phosphate solutions will give a good index of the available S in the soil because they will extract soluble sulfates, adsorbed sulfates, and some fraction of easily extractable organic S. Efficiency of the soil testing methods can be increased if care is taken in sampling the soil, processing the sample, analyzing the extract, and extrapolating the results. Knowledge of soil taxonomy should be used in interpreting the soil analysis. Tissue testing can further improve the interpretation of results and better delineate the S-deficient soils. In the tropical countries very little work has been done in correlating the results in the same taxonomic group of soils, and this makes the extrapolation of results all the more difficult.

Determinants of Sulfur Deficiency

Sulfur deficiency can be chronic, transient, and induced, depending upon the soil, climate, crop, and management system. To identify the S-deficient areas and to understand what causes or accentuates these deficiencies, present and future, there is a need to analyze (1) soil factors, (2) climatic factors, (3) crop, cropping system, and crop management, (4) fertilizer use and management, (5) irrigation, and (6) industrialization and environmental policies.

Soil Factors

The data available on the S status of the virgin tropical soils, specifically those relating the S content to the soil taxonomic group, are limited. In recent years, however, efforts have been made to determine the amount, form, and availability of S in different soil taxonomic groups. Generally the soils derived from volcanic parent materials, as are common in Central and South America, are deficient in S (Fitts, 1970). In these soils the organic matter is closely associated with the allophane, and the rate of release of S from this material is rather low. Despite their high organic matter contents, such soils are deficient in S, and crops planted on them respond to S application.

The coarse-textured soils in the humid as well as semiarid tropics (Table 5.2) are generally inherently poor in S and also have low retention capacity for sulfate added from various sources. Widespread S deficiency has been observed in light-textured soils in Asia and Africa, and such deficiencies become more evident in oilseed crops, legumes, forages, cotton, and cereals. Judging from the responses to S reported, it appears that these sandy soils of the semiarid tropics, which are agriculturally very important, may be vulnerable to S deficiency because of their inherently low organic matter, the high temperatures, and their high permeability. Kanwar (1963) observed that 75% of the groundnut-growing soils of Punjab are deficient in S and contain less than 10 ppm of extractable sulfate S. Approximately, 50% of these are also deficient in P, which makes the use of SSP an ideal fertilizer for correcting both S and P deficiencies. For correcting S deficiency, however, gypsum is adequate.

Little work has been done in the tropics to relate S deficiencies to soil factors, and the experience from the temperate zone is not very helpful. Furthermore, its extrapolation to tropical conditions is misleading. However, the relationship that has been observed in comparable tropical areas of Queensland and New South Wales in Australia or Hawaii in the United States could be extrapolated to other tropical conditions. For instance, Spencer (1966) observed that the color of basaltic sedentary soils in northern New South Wales was correlated with S status, the reddest being most deficient. However, there was no relationship between pH or redox potential and S deficiencies. The red soils, probably because of the high amounts of iron oxide, better aeration, and drainage, show greater oxidation of organic S and higher leaching losses. One could extrapolate this information to Alfisols which have been observed to be deficient in S in Africa, particularly in Nigeria. The research on S in west Africa provides good evidence of this type of relationship.

Attempts have also been made to relate S deficiency to geochemical information about soils, but the information is so scanty that no generalized pattern could be conceived. This makes a strong case for more research studies of associations of S responses to soil factors. Andrew (1975) observed that soils with a mean S content of 130 ppm responded to S application in Queensland, a tropical area. A large number of soils in the semiarid tropics have even less than this amount of S and are likely to benefit from S application.

Soil erosion can seriously affect the S status of soil. Data on S losses due to erosion are not readily available, but one can estimate the losses from the data on loss of organic matter and sulfate in the runoff water or windblown material. Although these losses may be serious in some locations, the eroded material whenever deposited can enrich the soil in S. Examples of this type of redistribution are available in all of the semiarid tropics and humid tropics. For soils that do not adsorb sulfate-S or sorb it weakly, there are only two sources of S for crops — mineralized organic S and atmospheric S. Any surplus sulfate will be leached out.

In a field experiment on hybrid maize at the Indian Agricultural Research Institute (IARI), New Delhi, Das et al. (1975) observed that the N response curves in the range 0-80 kg N did not show any effect of the addition of 30 kg S. However, beyond that it had a positive effect which increased as the N dose increased (Figure 5.3). The soil was highly deficient in N, but it contained a medium amount of available S (extractable sulfate-S > 10 ppm and total S even more than 300 ppm). It seems that S supply from the soil became a limiting factor only at high levels of N. Similar N response curves have been reported by Beaton (1980) which show the synergistic effect of S on response to N. These data indicate that S becomes a limiting factor at high levels of N use but the amount of S in the soil is probably adequate at low levels of yields and N doses.

Climatic Factors

Of the climatic factors rainfall, temperature, and soil moisture have the most relevance for S. Hasan, Fox, and Boyd (1970) concluded that S accumulation and distribution in soil profiles are related to rainfall in three

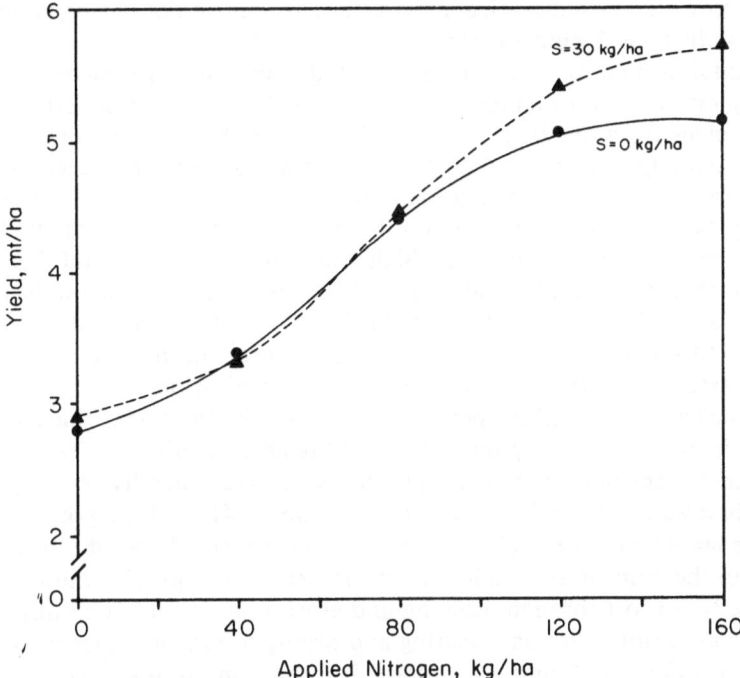

Figure 5.3. Effect of Applied Sulfur and Nitrogen on Maize Yield in New Delhi, India.

ways: (1) the capacity of soils to stabilize organic matter containing S is a function of reactive products of weathering which is related to rainfall, (2) the capacity of soils to adsorb sulfate is related to weathering products, and (3) rainfall is a major source of sulfate extracted from the air. The Australian studies in New South Wales show that, although rainfall and sulfate adsorption capacity have been reated, the relationship varied between soil types and no direct relationship was found to exist between adsorption capacity and S-supplying power of soils (Blair and Nicolson, 1975).

Fox (1980) has drawn attention to a number of reviews which indicate that S deficiencies are most pronounced in highly weathered tropical soils. These are also the soils with high sulfate adsorption capacity. For example, 58% of readily available S in a group of soils from Brazil came from adsorbed sulfate, whereas hardly 2% was contributed by this source in Iowa soil (Neptune, Tabatabai, and Hanway, 1975). It has also been observed that, though highly weathered tropical soils may contain thousands of kilograms of adsorbed sulfate per hectare even within the root zone of the crop, the crops planted in this soil may still respond to application of sulfate. This is because the amount of sulfate entering solution and being used by the plant roots may not be adequate to meet the plant needs. From numerous studies reported, Fox (1980) concluded that approximately 5 ppm of sulfate-S in the soil solution appears to be the critical limit for most of the crops in the tropical and subtropical regions though some crops like bananas may need as little as 2 ppm sulfate-S.

Rainfall is a major source of S extracted from the atmosphere. Coleman (1966), on the basis of a literature review, concluded that rainfall deposition of S ranges from less than 1.1 kg/ha to over 134 kg/ha. Low values are in rural areas. In the industrialized countries of the temperate regions large transfers of S from the atmosphere to the soil have been observed. However, in tropical countries the addition of S from the atmosphere is small. Sulfur inputs from rainfall are generally higher near the sea than inland. Fox et al. (1965) recorded 5 ppm S in rainwater 0.5 km away from the sea on the island of Kawai in Hawaii but only 0.8 ppm S, 8 km inland. The contribution of rainfall to soil S supply and the losses through drainage need to be quantitatively assessed to determine the need for S fertilization.

In the tropics the soil temperatures are generally very high, which accelerates mineralization of organic matter and release of sulfate-S. This process is further accelerated during dry periods which are generally very long(5-10 months a year), especially in the semiarid tropics. Thus, during the dry period the large amount of sulfate in the soil may cause a flush of S to be supplied to the crops in the beginning of the wet season, provided most of the sulfate is not lost through leaching and erosion during intense rains of the monsoon season. Alternate wetting and drying results in higher than usual mineralization and buildup of sulfate. Information on this subject is virtually nonexistent; there is an urgent need for studies on seasonal changes in the sulfate status of soil since such studies would hold the key to rational S fertilization.

Crops, Cropping Systems, and Crop Residue

The S requirements of crops depend on the nature and variety of the crop, the cropping system, and the expected yield. There are large variations in the amount of S removed from soils by different important crops of the tropical countries. While S needs may have been low at low crop yields, the use of highyielding varieties responsive to higher levels of nitrogenous fertilizers, will further increase the demands for S. Thus the areas that have shown S deficiency in the past may need fertilization with S-containing fertilizers, and other areas where S deficiency was either undetected or marginal will become more obviously deficient with higher demands for other nutrients.

The net amount of S removed by a crop is governed by the amount of S removed in the product and the fate of the plant residue. Because of nutrient recycling and generally low growth rate of native vegetation, the S requirement of these systems is generally low; when this vegetation is replaced by an agriculture crop, as is happening in many tropical countries of Africa and Latin America, the demand for S increases considerably. The increased S demand puts stress on the supply of S from the soil, and the soil may or may not be able to meet this demand. The changes in available S under an intensive cropping system in the sandy loam soil of New Delhi, as reported by Subbarao and Ghosh (1981), provide evidence of this process (Figure 5.4). The time period before S deficiencies are experienced varies depending on soil reserves, turnover rates, and the inputs from external sources.

It is not only a question of a particular crop variety but also of a cropping system and intensity of cropping. With higher intensity of cropping, sequential cropping, intercropping, relay cropping, and companion cropping, the demands on the S reserves in the soil will grow, and without adequate fertilization with S, it will no longer be possible to get good crop yields. Subsistence farming and shifting agriculture systems are being replaced with more intensive and market-oriented cropping systems that accentuate the requirements for all nutrients, including S. A specific example of the impact of a shift from a single crop using traditional technology to a double crop using modern technology on S requirements is provided in Table 3.4.

For comparison, let us assume that the crop production has increased from 2 mt/ha grain (subsistence farming) to 10 mt/ha (intensive farming). In the traditional farming system the grain and straw are generally consumed on the farm. In the case of modern farming systems, let us assume that all the grain is sold off the farm, rice straw is burnt, and wheat straw

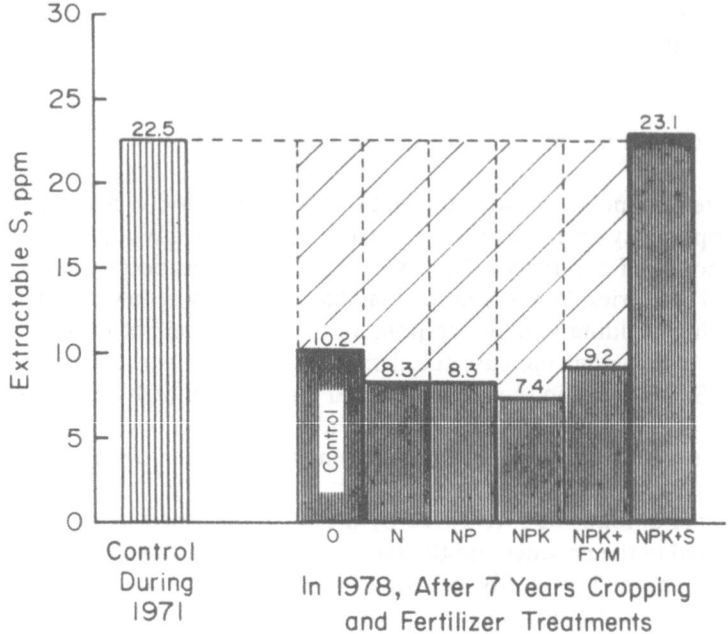

Figure 5.4. Changes in Sulfur Status Under Intensive Cropping and Different Fertilizer Treatments at IARI, New Delhi, India.

is recycled through the animal chain system.[1] In the intensive farming system, the estimated annual S loss for wheat would be 7.5 kg from straw and 5 kg from grain, making a total annual oss of 12.5 kg S/ha.[2] In the case of rice where straw is all burned and grain sold out, all the 15 kg S/ha will be lost annually. Thus the annual S loss from the farm will be 27.5 kg/ha. With subsistence farming in the same crop rotation and with low yields at one-fifth of this level, the annual loss would have been only 5. kg/ha. If only one crop were grown, the annual S loss would not exceed 2.75 kg/ha.

In case the annual accession of S from the atmosphere through rain is about 1.5 kg/ha, the net S deficit would be 1.25 kg/ha in one crop and 4.0

1. In the subsistence farming system, the annual yield is 2 mt/ha of grain and 3.32 mt/ha of straw; and total S removed is 7 kg/ha, of which 4.6 kg/ha is in straw. In the intensive farming system, the annual yield is 10 mt/ha (5 mt wheat and 5 mt rice) of grain, and 16.6 mt/ha (8.3 mt of wheat and 8.3 mt of rice) of straw; and the total S removed is 20 kg/ha of this 15 kg in straw) by wheat and 15 kg/ha (of this 8 kg in straw) by rice.
2. Based on Decau (1972): (a) 50% of S is returned to the soil when cereal grain is sold off the farm and straw is returned as manure, (b) 60% of S is returned to soil when cereal grain is consumed on the farm and manure is returned to the field, and (c) 70% of S is returned to the soil when forage is consumed on the farm and manure is returned to the field.

kg/ha in the two-crop subsistence farming; this could have been met from the soil and sulfate S in irrigation water and the farmyard manure. If the atmospheric S addition is 3 kg/ha/year, it would be even more favorable. But in the intensive farming system with an annual S deficit of 27.5 kg/ha, the supply of S from external sources such as S-containing fertilizers becomes essential. It has een assumed that the S supply from soil, the atmosphere, irrigation water, and armyard manures, is the same as in a subsistence farming system.

Moreover, there is a long history of nutrient export from this region without appropriate import or application of S to the soil. The removal of S by a fertilized crop of groundnuts (fertilized with 251 kg of SSP/ha) is 5.4 kg in haulms, 0.8 kg in shells, and 5.4 kg in kernels or a total of 11.6 kg S/ha. With 1.2 million ha under groundnuts in Nigeria this would amount to a loss of about 14,000 mt of S annually. A large part of this S may be exported through the export of groundnuts.

The Ludhiana soils are sandy with very little adsorption capacity for sulfate; thus, S deficiency is likely to be serious. It is no surprise that many recent studies from the Punjab Agriculture University, Ludhiana, have indicated the acute S deficiency and responses to S (Pasricha and Randhawa, 1973; Kanwar, 1963; Dalal, Kanwar, and Saini, 1963; Kanwar and Randhawa, 1974; Chopra and Kanwar, 1966; Aulakh, Pasricha, and Dev, 1977; Dev and Kumar, 1982). Bromfield, Hancock, and Debenham (1982) have cited similar cases of increasing S deficiency in the highlands of Kenya. The long-term studies that were started by the Indian Council of Agricultural Research in India in 1971 also bring out convincingly the increasing need for S (Ghosh, 1980). Crops, including wheat, groundnuts, Indian mustard and rape, berseem, chickpeas, and maize, have indicated good responses to S application in these soils. The use of gypsum has become popular in this state, not only for reclamation of sodic soils, but also for correcting S deficiency in normal soils that are either inherently deficient or are experiencing induced deficiency due to higher yields, intensive cropping, and higher use of S-free NPK fertilizers.

The S removal by the product depends on both the S content of the product and its yield (Masters and McCance, 1939). It should be pointed out that the literature shows great variation in S content of grains and other products (Table 4.2). For example, S in rice grain may vary from 0.034% under S deficiency to 0.16% under S sufficiency, and the yield of grain may vary from 0.75 mt to 8.0 mt of grain. Sulfur requirements will vary from 0.26 to 12.8 kg/ha (Blair, 1979). Another important factor in considering the S supply and removal is the rooting depth and zone of exploitation by the crop. The short-rooted forages and annual crops exploiting the surface soil may suffer from acute S deficiency, whereas deep-rooted perennial crops in the same soil may be able to use the S adsorbed or accumulated in lower horizons of the soil and may not show S deficiency.

82

An evidence of increasing S deficiency in Sudanese savanna > Guinea savanna > derived savanna > forest zone of the humid tropics of Nigeria is provided by Kang et al. (1981). The losses of S through loss of organic matter, through mineralization, or through burning of vegetation or clearing the forests and savanna woodlands, have already been discussed. It may, however, not be out of place to emphasize that with shifting cultivation and immediately after clearing the forests, the deficiencies of S in tropical areas become more manifest. Bromfield et al. (1982), on the basis of their experience in Nigeria and Kenya, have postulated the following (Figure 5.5):

Stage 1 — In the first year after clearing of natural vegetation or cultivation of fallow land or old pasture land, less organic S is mineralized because more organic matter is incorporated in the soil. This may cause a temporary or transient S deficiency. The case is similar to N deficiency occurring after clearance of vegetation.

Stage 2 — After 2 years the amount of S released from mineralization of organic matter and other sources of S provides enough S for the crop. Now the transient S deficiency disappears and, in fact, an excess of S occurs.

Stage 3 — With continued cropping the combined sulfate-S supplies exceed the crop requirement, and the surplus sulfate moves into the subsoil and gets adsorbed. The depths of movement and adsorption depend on the nature of the soil, the amount of water leaching the sulfate, and the competition of the crop for removing it. These conditions may lead to a stage of chronic deficiency.

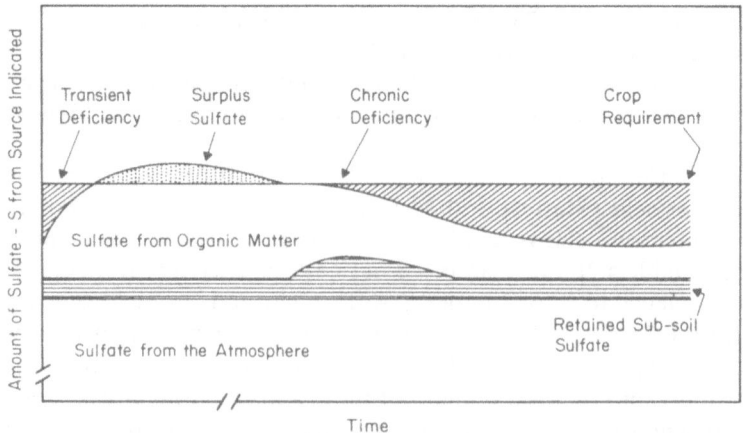

Figure 5.5. Sulfur Deficiency and Changes in Sources of Sulfur to Crops in Soils From Clearance of Natural Fallow Through Cropping Cycle With No Application of Fertilizer Sulfur (not to scale).

Stage 4 — With continued cropping the organic S and the amounts of sulfate-S released by mineralization decline. The adsorbed S also declines, the total S supply to plant becomes less, and a perennial S deficiency appears, which may be called induced deficiency. Bromfield et al. (1982) called it chronic deficiency, but we would prefer to call it induced deficiency.

If the crop and soil management systems are based on incorporation of residues of the previous crop directly into the soil, the S deficiency may not become so acute. On most of the farms in the developing countries, particularly in the tropical zone, the straw is mostly removed from the system either through burning or by export, and this accentuates S deficiency problems. In Punjab (India) rice straw is being burned, which means that more than three-fourths of its S is lost. Besides creating an environmental pollution problem and a health hazard, this creates an S deficiency, which will be aggravated with more intensive cropping and a continuance of the practice. At present, Punjab is reported to be burning more than 5 million mt of straw in the fields.

The beneficial effect of straw will depend on the amount of S incorporated in it. In some instances straw may even decrese the amount of S available to plants by immobilizing it through bacterial action. Stewart, Porter, and Viets (1966) found there was a net immobilization of soil S when wheat straw with less than 0.15% S was added to the soil. Thus application of S may even improve the efficiency of S provided through crop residues and straws.

Fertilizer Use and Management

Fertilizer Type — One of the primary external sources of S supply is S-containing fertilizers. The low proportion of S-containing materials and the use of high-analysis N, P, and K fertilizers, which usually have little or no S, are creating S deficiencies in modern agriculture. Hignett (1970, 1974) has drawn attention to this reduction in S input due to the shift from AS to urea, and from SSP to TSP.

With proper selection and use of S-containing fertilizer, the S needs of a crop can be met. If this is not feasible, the use of materials like gypsum, phosphogypsum, pyrites, and elemental S has to be considered. There is also a need for the development of technology for the manufacture of fertilizer products that would supply other nutrients and provide an inexpensive source of S. For economic reasons TSP, which has 46% P_2O_5 and no S, has replaced SSP, which has 16%-18% P_2O_5 and 12% S. Trends also indicate greater use of TSP in the future and thus aggravation of the S deficiency. Greater use of DAP and phosphate rock also would lead to greater imbalance between P and S supply.

In greenhouse and laboratory studies, Korentajer, Byrnes, and Hellums

(1983) observed that liming of acid soils increases the leaching losses of sulfate due to several mechanisms such as desorption of sulfate from soil colloids, increased mineralization, and increased solubility of sparingly soluble hydroxysulfate compounds. The relative effects of these factors depend on the soil profiles, the amounts of organic S and adsorbed sulfate, the sulfate adsorption capacity of the soil, the soil pH, and the temperature. Pertinent data from their experiments for two soils are given in Table 5.9. Both soils tested belonged to the same taxonomic group and were highly acidic. Mountview soil had higher cation exchange capacity (9.03 meq/100 g) and Hartsells lower (5.08 meq/100 g). These results show that in the absence of leaching liming did not materially change the sulfate content of the soil that was extractable with water or phosphate, but with continuous leaching there was a marked decrease in sulfate-S extractable with phosphate or water. The change was more marked as the level of liming increased, and it was also greater in the soil with greater sulfate adsorption capacity.

According to these studies, when acid soils that are potentially S deficient are limed and subsequently leached — which would happen under high rainfall conditions — they may become S deficient, and expected effects of

Table 5.9. The effect of liming and leaching on soil sulfate, total soil sulfur, and the amounts of sulfate leached[a].

Liming rate (g/kg)	pH	H_2O-extractable SO_4-S (μg S/g)		H_2PO_4-extractable SO_4-S (μg S/g)		Adsorbed SO_4-S[c] (μg S/g)		Total Sulfur (μg S/g)		Leaching loss (μg S/g)
		− [b]	+	−	+	−	+	−	+	+
Mountview soil										
0.0	4.3	14.4	9.5	14.2	9.5	0	0	139	114	15
0.5	4.7	15.6	2.0	18.1	1.3	2.5	<0	d	d	19
1.0	5.0	14.0	0.8	18.9	1.3	4.9	0.5	154	125	24
1.5	5.4	18.5	0.6	20.0	2.1	1.5	1.5	149	105	29
2.0	5.6	20.5	0.5	21.7	1.3	1.2	0.8	151	89	34
Hartsells subsoil										
0.0	4.5	6.6	3.4	50.7	24.1	44.1	20.7	107	65	41
1.0	6.3	18.9	1.3	53.1	3.2	34.2	1.9	102	57	64
1.8	6.8	29.8	1.4	75.9	4.5	46.1	3.1	98	60	72
2.5	7.0	33.3	1.8	84.2	4.8	50.9	3.0	107	37	71

a. After 2-week liming and incubation at field moisture capacity (20 °C) and the 10-day leaching period.
b. The '−' means not leached, '+' means leached.
c. Calculated as adsorbed SO_4 = (H_2PO_4-extractable SO_4) − (H_2O-extractable SO_4).
d. Not available.
Source: Korentajer, Byrnes, and Hellums (1983).

liming in increasing yield may not be realized because of lime-induced S deficiency. Of course, it is also true that many deficiencies in micronutrients such as Zn, Mn, Fe, and B also become more marked on liming, but liming-induced S deficiency has been more often overlooked in the past. In the highly weathered tropical soils this may be a serious limitation to crop production.

Intensive Use of Phosphates – Phosphate fixation and phosphate deficiency in the tropical soils are well known, and any program for crop production aims at regular and heavy use of phosphates. In fact, efforts are made to saturate the phosphate-fixing capacity of these soils with heavy doses of phosphates, which indirectly reduces the possibility of sulfate adsorption and thereby induces rapid losses of sulfate through percolating water under torrential tropical rains.

The intensive use of phosphates, particularly those that are free of sulfate, will aggravate S deficiency in soils by creating an imbalance of S and P, which will cause the phosphate to displace the sulfate from the adsorption surfaces. Ensminger (1954) reported that after 18 years of application of P and S, the soil with the highest rates of P and S was lowest in extractable sulfate-S. The sulfate thus displaced may go into solution and be taken up by the growing crop, or it may be leached out of the soil. Some sulfate ions moving with the percolating water may be adsorbed on the surface of the soil complex in the lower horizons of the soil.

Thus, overphosphating, like overliming without consideration of fate of sulfate in soils, is gradually making the tropical soils deficient in S. Lime and phosphate applications are no doubt essential for improving productivity of acid soils high in exchangeable aluminum. But to avoid the adverse effect of these applications on sulfate reserves in the soil, the use of coarse-grade gypsum or other sparingly soluble sulfate-containing fertilizers may be useful. It is also evident that the problems of S fertilization of the crop should not be studied without considering the interactions of lime, phosphate, and other nutrients in the field. It is doubtful that the single nutrient approach to fertilization is meaningful.

The waterlogged soils of hot and warm tropical regions are extensively used for growing rice. The effect of the addition of lime, AS, and SSP in waterlogged rice soils of Assam was studied by Haldhar and Barthakur (1976). They observed that these treatments increased the formation of hydrogen sulfide and water-soluble sulfides in the soil. Thus the reduction of sulfate to S, encouraged by waterlogging and treatments with lime, SSP, and AS, caused sulfide injury to growing rice plants. Sachdev and Chhabra (1974) reported that after 4 months of incubation with sulfate, under anaerobic conditions in an alkaline soil, barely 12.1% of the added S could be detected as sulfate-S, whereas under aerobic conditions 68% was still present in sulfate form. This represents another type of S loss in lowland paddies.

Thus the problem of fertilization and management of fertility through judicious use of lime and phosphate has practical implication for S availability in tropical soils, but such studies should be made under the field environments rather than the laboratory conditions.

Irrigation

Irrigation in arid and semiarid tropics can have two types of effects: (1) it can add sulfate to the soil and (2) it can leach the sulfate from the soil. Depending upon the quality of water, the concentration of sulfate, and the nature of the soil and crop management systems, it will have different effects. The lighttextured sandy soils of the semiarid tropics in Africa and India have shown considerable S deficiency and hence responses to S. It is interesting that despite application of irrigation water which may contain considerable amounts of sulfate (1-5 meq/liter) these soils still respond to applications of S. Because of the high rate of movement of water through the soil and the low retention capacity of soils, there is need for regular application of S.

Thus irrigation water itself acts as a leaching medium for the sulfate from the soil and may aggravate S deficiency. Under monsoon rainfall conditions, particularly if there is continuous rainfall for some time, S deficiency, like the N deficiency, is likely to increase. Attention has been paid to leaching losses of nitrates from soils, but no attention has been paid to reducing the leaching losses of sulfate. Split application of N fertilizers is advocated for reducing N losses. Split application of sulfate may prove equally beneficial and also deserves consideration. Use of sulfur-coated nitrogen, potassium sulfate, and sulfur-fortified phosphates needs examination.

Korentajer, Byrnes, and Hellums (1983) studied the effect of leaching on applied S. These authors used 0.87, 1.78, and 4.89 mm leaching/day and S application rates equivalent to 0, 10, 20, 40 kg/ha. They observed that with the 10-kg S application rate and the leaching rate of 4.89 mm/day, which would be considered normal for coarse-textured tropical soils, 44% of the S was lost in leaching and 43% was recovered by the plants. The rest (23%) was retained in the soil. The authors concluded that moderate-to-high percolation rates, such as those encountered in many S-deficient soils of tropical regions, will lead to significant losses of sulfate and decreased fertilizer efficiency. Rhue and Kamprath (1973) indicated the changes in sulfate status in the A horizon of an Arenic Paleudult at various times after application of gypsum (Table 5.10).

Yoshida and Chaudhry (1979) suggested that irrigation water with an S content of 2.7 ppm should be able to supply the entire S needs of a rice crop, whereas Wang (1978) considered 6 ppm to be the minimum required. Responses to S were experienced in rice paddies in South Sulawesi, Indonesia, despite the sulfate-S content of irrigation water being more than 2.8

Table 5.10. Changes in soluble sulfate in soil at various depths and different periods.

Days after treatment with gypsum	Soluble SO$_4$-S content[a]		
	0 – 15 cm (ppm)	15 – 30 cm (ppm)	30 – 45 cm (ppm)
44	22	20	8
91	14	14	18
154	2	7	7
188	1	2	4

a. Initital SO$_4$ at all depths was less than 1.4 ppm.
Source: Rhue and Kamprath (1973).

ppm. Indian experience suggests that many of the waters from the tubewells and canals, which are commonly used for irrigation in Punjab, have more than 6 ppm of sulfate. Yet the crops irrigated with these waters respond to S application. Yadav (1982) gives the mean chemical composition of ground waters of Punjab and Rajasthan, which are used extensively for irrigation (Table 5.11). A survey of Indian literature suggests that even in these districts, despite the high sulfate-S content of their soils, responses to S are common, particularly when these waters are used on sandy or light-textured soils.

The canal water in Northern India has total salt content of less than 160 ppm, and its S content does not exceed 10-12 ppm. In a recent survey of ground waters used for irrigation in the Sangrur district in Punjab, Singh

Table 5.11. Mean electrical conductivity (EC) and chemical composition of some ground waters of Punjab and Rajasthan in India.

State	EC $\times 10^3$ (micromhos/cm)	Salt content (approximate) (ppm)	SO$_4$ content	
			meq/liter	S (ppm)
Punjab				
Ferozepur	2.1	1 344	5.8	92.8
Bhatinda	2.7	1 728	5.0	80.0
Sangrur	1.5	960	10.6	169.6
Amritsar	1.0	640	0.5	8.0
Rajasthan				
Ajmer	5.5	3 520	24.8	396.8
Bikaner	6.3	4 032	12.2	195.2
Jodhpur	5.5	3 520	8.7	139.2
Pali	5.1	3 264	10.8	172.8

Source: Yadav (1982).

88

and Marok (1980) observed that, of 446 water samples, 84% contained between 3.2 and 72 ppm sulfate-S. Aulakh and Dev (1976) have reported S deficiency in this district. The S needs of the crops on these soils have not yet been studied. There is a need to examine the role of sulfate added through irrigation water on crop nutrition. It is possible that in coarse-textured soils of high permeability, low organic matter content, and low capacity for adsorbing sulfate, the sulfate from irrigation water goes eyond the reach of the plant. Or perhaps some chemical changes, such as precipitation, make the sulfate unavailable.

Acute cases of S deficiency in rice have been reported in Indonesia and Bangladesh, especially in areas where irrigation water is low in sulfate-S. Ismunadji and Zulkarnaini (1978) and Sakai (1980) have reported the S content of irrigation waters where rice responds to S application. The data in Table 5.12 show that nearly all samples had very low sulfate-S content. It seems that the relationship between the sulfate content of irrigation water

Table 5.12. Sulfur content of irrigation water in Indonesia and Bangladesh.

Country	Location	S content (ppm)
Indonesia[a]	Muara Bogor	1.28
	Citayam Bogor	1.36
	Singamerta Serang	4.04
	Cihea Cianjur	2.64
	Megalang	6.17
	Meguwahargo Yogjakarta	6.17
	Ngale Ngawi	1.68
	Pacet	20.20
	Pusakanegara	19.40
Bangladesh[b]	Joydebpur[c]	0.06
	Dhaka roadside[d]	1.22 – 2.78
	Comilla substation[d]	0.99
	Bangladesh Agricultural Research Institute (BARI) farm[e]	0.20
	Joydebpur[e]	0.26 – 0.27
	BARI Aman[e]	0.13 – 0.54
	BARI West Boro[f]	7.18
	Deep water Aus[f]	3.58
	BARI main drainage canal[f]	7.38
	Bansi[g]	3.02
	Jamna[g]	4.59

a. From Ismunadji and Zulkarnaini (1978).
b. From Sakai (1980).
c. Rain water.
d. Tank water.
e. Tubewell water.
f. Paddy field water.
g. River water.

and the S deficiency in soils irrigated with this water needs to be thoroughly investigated; some irrigation waters have reasonably high sulfate contents, and yet the soils irrigated with them still respond to S. The problem has practical significance since in all developing countries irrigation is being given high priority for increasing food production. If irrigation water can meet the S needs of the crop, the need for S fertilizers will diminish.

If the soils are inherently deficient in S, rainfed agriculture will experience an S deficiency earlier than will irrigated agriculture partly because it lacks the inadvertent addition of sulfate from irrigation water. Experience with groundnuts from Asia and Africa confirms this view, since groundnuts are mostly grown under nonirrigated conditions.

Industrialization and Environmental Policies

The effect of industrialization upon S deficiency in the soil is primarily that industry emits SO_2 to the atmosphere, and this becomes a source of S to the soil. However, the accession of S through this mechanism depends on the nature of the industry, the governmental policies regarding clean air, the distance of the industry from the cropped area, the vegetative cover, and crop management system.

In tropical countries the level of industrialization is generally rather low. Therefore, the contribution of industry to S status of soils is not an important factor. However, in the vicinity of the industrial units it can become an important source of S. The effect of stringent measures being adopted by industry for clean air may also lead to reduction in accession of S to soils. A detailed analysis of the case of S deficiency being created in Zambia by such environmental control measures indicates the S deficiency problems that agriculture will face under such conditions (Kiyoura, 1982). There are examples of higher crop yields near the copper smelting plants in Zambia. It thus follows that in the formulation of environmental policies, the beneficial effect of SO_2 should be given due consideration.

Conclusions and Research Agenda

1. There is a dearth of information about the amounts, forms, and distribution of S in the soils of the tropical regions, particularly in relation to typical taxonomic soil groups under different ecosystems and the changesthat occur under different management systems.
2. Even the meager soil S data available indicate that tropical soils are inherently poor in S supply and may become even more impoverished under intensive cropping if proper care is not taken to replenish the S removed by crops.
3. There is a need for standardization of techniques for evaluation of S sta-

tus of the soils, particularly for determination of available S for diagnosis of S deficiency and S fertilizer needs of the soil for optimum crop production.

4. Extraction of soil S with monocalcium phosphate solutions seems to offer greater promise for delineation of S-deficient soils, but for relating the responses to fertilizer S, a combination of soil test and plant analysis is desirable. There is a need for establishing critical limits of available S for different crops and soils. Note should be taken of the relationship of sulfate-S in tissue to total S as an index of S needs.

5. Limitations of soil tests have been recognized by many researchers. More emphasis should be placed on field experimentation and soil and plant analyses to correlate the responses to S in the real world situation.

6. The laboratory techniques for estimation of S from the soil extract or plant tissue need to be improved, and instrumental analytical techniques such as spectrographic techniques should be standardized.

7. Contribution of the subsoil adsorbed S to crop nutrition should be given due consideration.

8. Because SO_2 from the atmosphere and sulfate-S from the irrigation water modify the S needs, due care should be taken in conducting greenhouse studies; they should be limited to conditions in which exclusion of these amounts of S would not limit the utility of the results.

9. For comparison of S sources, greenhouse studies on a soil representative of the region can be conducted, but the limitations of the extrapolation of these results to field conditions should be recognized. Thus it becomes necessary to verify the conclusions derived from the greenhouse studies under the actual field conditions.

6 Crop Response to Fertilizer Sulfur in the Tropics

The soils in the tropical countries of Asia, Africa, and Latin America are facing widespread S deficiency. In many cases S deficiency is localized or is specific to a crop, soil system, or an agroclimatic region. The application of sulfur on S-deficient soils generally results in positive crop response. The primary purpose of this chapter is to evaluate the nature and magnitude of crop responses to S fertilization in tropical developing countries of Asia, Africa, and Latin America.

Asia

In recent years S deficiency in many food crops has engaged the attention of scientists in South Asia, Southeast Asia, and East Asia. Sulfur studies have been conducted in many countries in these regions. The crop response to applied S will be discussed separately for each of the countries.

India

Sulfur deficiencies in Indian soils have been reviewed by Kanwar and Randhawa (1974) and more recently by Dev and Kumar (1982). Sulfur deficiencies are widespread in Punjab, Haryana, Himachal Pradesh, Uttar Pradesh, Rajasthan, Bihar, West Bengal, and many areas of southern India. The light-textured soils − particularly alluvial (Entisol, Inceptisol), coastal (alluvial), laterites (Oxisol), and red (Alfisols) − and even black soils (Vertisols) have been reported to be deficient in S (Naik and Das, 1964). The S deficiencies have been reported in groundnuts, rapeseed, mustard, tea, coffee, sugarcane, jute, chickpeas, mung beans, soybeans, wheat, maize, sorghum, rice, and forages such as berseem and alfalfa.

Wheat − Aulakh, Pasricha, and Dev (1977) reported that in Punjab the application of 25 kg S/ha increased the grain yield of wheat by 480, 553, and 888 kg/ha in K 227, PV 18, and S 308 varieties of wheat, respectively. Differential varietal responses to S have also been reported by Pasricha, Bajwa, and Randhawa (1975) (Figure 6.1). Joshi and Seth (1975) found that 50 kg S/ha was optimum for wheat in Rajasthan but with a higher dose of phosphate (100 kg P_2O_5/ha as TSP), the S requirement increased to 75 kg S/ha. As reported in Dev and Kumar (1982), Ruhal obtained the highest yield of wheat with 80 kg S/ha in Rajasthan. Marok (1978) reported that in Ferozepur district of Punjab the PV 18 variety of wheat gave 1,606 and 1,840 kg/ha more yield with SSP than with DAP when compared on equivalent P and N basis with urea or CAN, respectively, as sources of N. The difference was attributed to S added through SSP. In a recent study by

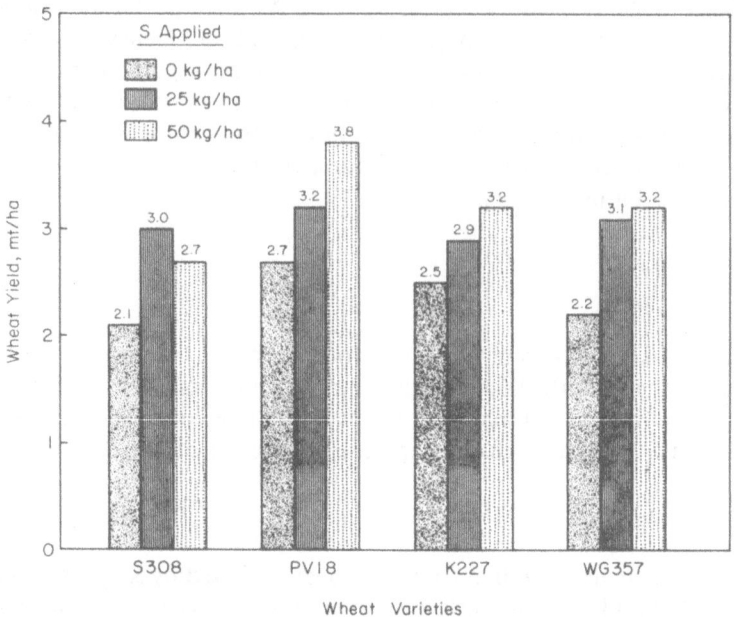

Figure 6.1. Effect of Applied Sulfur (Gypsum) on High-Yielding Wheat Varieties in Light-Textured Soils in Ludhiana, Punjab, India, in 1971–73 (2-Year Mean Yields at Constant Levels of N, P, and K).

Arora et al. (1983) on wheat that followed groundnuts, typical symptoms of S deficiency appeared on farmers' fields in Ludhiana district of Punjab. An application of S from gypsum, pyrite, or AS to a crop 40-50 days old, produced a spectacular effect on the crop, and the yield increase varied in the following order: ammonium sulfate > pyrites > gypsum.

Maize – Ruhal found that application of 90 kg S/ha significantly raised the yield of maize in S-deficient soils of Rajasthan (Dev and Kumar, 1982). Pasricha et al. (1977) and Dev, Jaggi, and Aulakh (1979) reported significant effects of S application on maize yields in the Punjab. In their experiment 20-25 ppm of S applied through fertilizers produced the optimum yield.

Rice – A classical example of S deficiency in rice was reported by Aiyar (1945). Some uncoordinated work on S responses in rice has been reported by a few workers (Acharya, 1973) and, in many cases, the apparent superiority of AS over other N sources and of SSP over TSP has been attributed to the S content.

Groundnuts – Kanwar (1963) reported that 75% of the area planted to groundnuts in Punjab was deficient in S, and 50% was deficient in both S and P. Dalal, Kanwar, and Saini (1963) observed that the yield of ground- nuts was increased over the control by 34% with the application of AS, by 46% with SSP,and by 41% with gypsum (Table 6.1). These responses were much greater than could be attributed to the N, P, and Ca content of these fertilizers. The soil was extremely deficient in P and S. Pathak and Pathak (1972) in Uttar Pradesh and Ruhal in Rajasthan (Dev and Kumar, 1982) reported the highest yield of groundnuts with fertilizer treatments contain- ing S. Aulakh, Pasricha, and Dev (1977) observed that 32 kg S/ha was enough for obtaining optimum yield of groundnuts in the Punjab. Aulakh, Pasricha, and Sahota (1980a) also reported that in a comparison of three sources of phosphates (SSP, TSP, and DAP), SSP produced significantly higher yields of groundnuts than did the others (Figure 6.2). However, the differences disappeared when equivalent amounts of S were added to the other phosphate treatments. From a review of results of field experiments from all of India, Kanwar, Nijhawan, and Raheja (1983) concluded that op- timum yields of groundnuts could be obtained by application of S in the form of gypsum.

Mustard − The unfavorable effect of S deficiency on grain yield and oil content has been reported from most of the areas growing mustard in Pun- jab, Haryana, Rajasthan, and Uttar Pradesh. More scientific papers have been published on this crop in the last decade than on any other oilseed crop (Singh and Moolani, 1970). Singh andSingh (1977) reported a linear response in grain yield with a dose up to 60 ppm S, whereas Pasricha and Randhawa (1973) found that application of 25 ppm S was adequate for ob- taining optimum yield of mustard in Punjab. Aulakh, Pasricha, and Sahota

Table 6.1 Effect of different sulfur-containing and sulfur-free fertilizer treatments on groundnuts in sandy loam soil of Ludhiana, Punjab, India[a].

Treatment	S content of fertilizer added (kg/ha)	Mean response over control[b] (%)
Ammonium sulfate	32 kg	34
Ammonium chloride	–	21
Single superphosphate	32 kg	46
Triple superphosphate	–	32
Gypsum	32 kg	41
Calcium chloride	–	26

a. Nitrogenous fertilizers were equated on N basis, phosphates on P basis, and calcium salts on Ca basis.
b. Based on 3-year (1959 – 61) average pod yield. The mean pod yield for the control was 1 176 kg/ha.
Source: Dalal, Kanwar, and Saini (1963).

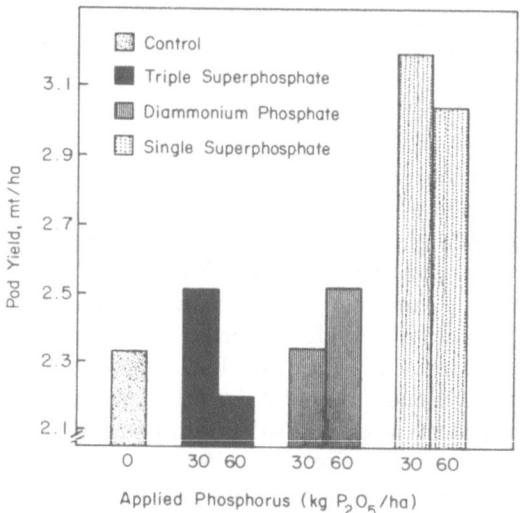

Figure 6.2. Effect of Single Superphosphate in Comparison with Triple Superphosphate and Diammonium Phosphate on Pod Yield of Groundnuts.

(1977) reported that maximum grain yield was obtained with 30 kg S/ha supplied as gypsum, along with 120 kg N as urea. Even 20 kg S, along with N, increased yield of RL18, RLM198, and RLM154 varieties of mustard by 155%, 167%, and 180% over the control in the Punjab. The effect of S, as obtained by Dev, Saggar, and Bajwa (1981), on the grain yield of different varieties of mustard in Punjab was good only up to 20 kg/ha of S. Beyond this, either there was no response or response was very small. The average mustard yield response to 20 kg/ha of S application was 25, 24, and 35 kg/1 kg of S for three different mustard varieties.

Aulakh, Pasricha, and Sahota (1980b) reported the results of 3 years of field experiments on yellow mustard (Brassica compastris L.) and Indian mustard (Brassica juncea L.). The soil was loamy sand with pH 8.9 and was deficient in S. Sources of S, N, and P were gypsum, urea, and TSP. The results (Table 6.2) show that maximum yields of grain and oil were obtained when both N and S rates were high, which indicates significant N x S interaction. The combined application of N and S had the largest effect on the concentration and uptake of N and S and on protein and oil content of grains and their yield. The authors concluded that 60 kg S with adequate N supply is likely to improve the yield and quality of mustard oil. Singh and Singh (1978) recommended a dose of 250 kg S/ha for obtaining the good yield of mustard, whereas Ruhal and Nad, as reported in Dev and Kumar (1982), found that 90 kg S was adequate. These experiments were conducted in Haryana, Rajasthan, and New Delhi, respectively.

Table 6.2. Effect of nitrogen and sulfur rates in trials at Samrala, Punjab, India.

Treatments[a]	N:S ratio[b]	Protein[b] (%)	Oil content[b] (%)	Grain yield[b] (kg/ha)	Oil yield[b] (kg/ha)
		Yellow mustard			
N_1S_1	9:6	14.6	40.9	230	100
N_4S_1	10:7	23.3	41.5	733	300
N_4S_2	10:3	27.5	44.4	707	310
N_4S_3	7:4	28.1	47.1	856	450
		Indian mustard			
N_1S_1	11:1	18.5	35.9	503	180
N_4S_1	10:2	22.5	36.4	1 486	540
N_4S_2	9:2	28.9	39.2	1 606	630
N_4S_3	6:9	30.8	42.6	1 570	670

a. S_1, S_2, and S_3 are 0, 30, 60 kg S/ha. $N_1 = 0$. N_4 is 75 kg N for yellow mustard and 180 kg for Indian mustard in the first 2 years and 90 kg N for yellow mustard and 150 kg for Indian mustard during the third year.
b. Three-year average.
Source: Aulakh, Pasricha, and Sahota, 1980b.

Soybeans − A level of 60 ppm S was reported to be optimum for getting the highest yield of soybeans (Pasricha and Randhawa, 1973). Subbiah and Singh(1970), Saggar and Dev (1974), and Dhillon and Dev (1978) also observed similar soybean response to S application.

Pulses − Aulakh and Pasricha (1979) observed that in a soil containing 16.8 kg/ha available S (which implies that soil is not S deficient), S application did not significantly increase the yield of chickpeas and lentils but did enhance the S content, which has favorable nutritional implications. The application of phosphate decreased the S content as well as yield. But with mung beans grown after chickpeas or lentils the residual effect of S on the yield as well as S content of the beans was positive. In another experiment Aulakh, Pasricha, and Dev (1977) observed that in a field deficient in S (8 ppm available S), 47% increase in yield of chickpeas, 27% in the lentils, and nearly 100% in residual effect on mung beans (planted after lentils) occurred with an application of 40 kg S/ha over and above 15 kg N/ha and 40 kg P_2O_5/ha (Figure 6.3). With black gram (also known as urd or mash) (Phaseolus mungo L.) the application of SSP, because of its S content, gave 110 and 152 kg/ha more grain yield than did DAP and TSP (Figure 6.3). An appreciable residual effect of S on mung beans grown after cowpeas and mustard was also reported by Nad (Dev ad Kumar, 1982).

Dube and Misra (1970) reported that in peas (Pisum sativum), black gram (Phaseolus mungo L.), chickpeas (Cicer arietinum L.), and groundnuts (Arachis hypogaea L.) S deficiency reduced the yield, quality, and protein

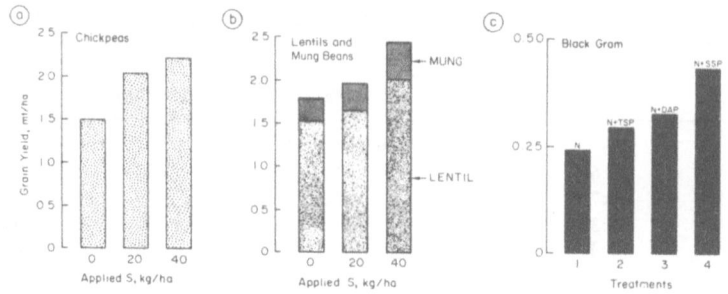

Figure 6.3. Effects of Applied Sulfur on Grain Yield of Pulses in Punjab, India.

content of the seed. Singh (1970) reported a 100% increase in yield of sweet peas by an application of 250 kg S/ha. Mehta and Singh (1979) observed that application of 250 kg S/ha to a calcareous soil of Rajasthan increased the yield of green gram by 95% and the chlorophyl content of leaves by 54%. The soil was very low in S.

Forages — Aulakh and Dev (1977) reported that alfalfa showed a significant response to up to 20 ppm of S in an S-deficient soil of Madhya Pradesh. Bansal, Sharma, and Singh (1979), on the other hand, found significant response to S up to 50 ppm in another soil. Pasricha and Randhawa (1975) observed that S applied as superphosphate and gypsum significantly increased the protein content of berseem in Punjab. Significant responses of berseem to S were also reported by Pathak and Bhardwaj (1968) in Uttar Pradesh and by Sisodia, Sawarkar, and Rai (1975) in Madhya Pradesh. Reddy and Mehta (1970b) reported that alfalfa showed a significant response to S application in loamy-sand soils of Gujarat.

Cotton, Sugarcane, and Jute — Dev and Kumar (1982) reported that a linear response in dry-matter yield of cotton was obtained with up to 60 ppm S but that beyond this amount, yield was reduced. Sulfur deficiency in sugarcane and jute was observed by Dutt (1962a,b). Saroha and Singh (1979) found that the use of S raised the yield of sugarcane in Vertisol soil in Udaipur, Rajasthan.

Tea — Tea is highly responsive to applications of S in the form of AS or SSP. Kanwar and Takkar (1966) reported an increase of 64.5% over control in yield of tea leaves in Kangra (Himachal Pradesh) by an application of AS supplying 196 kg N/ha and 224 kg S/ha. Similarly, because of significant responses to S and N in the tea-growing areas of Assam and Nilgris (South India), AS has become a preferred fertilizer for tea. Where AS is not available, urea and elemental S are used to supply N and S. Earlier studies by Ferguson, Gokhale, and Dutta (1957) and Child (1957) also bring out the need for S application in tea soils of Assam.

Sources of Sulfur — Most of the field experiments to correct S deficiency, in crops other than tea, have shown the superiority of gypsum as a source of S. Subbiah and Singh (1970) and Singh, Subbiah, and Gupta (1970) reported that gypsum was superior to AS and sodium sulfate on groundnuts and mustard, provided any N deficiency was remedied. For tea, however, AS proved superior to superphosphate (SSP), gypsum, and sodium sulfate (Kanwar and Takkar, 1966).

Saroha and Singh (1979) reported that the impact of S in improving the quality of sugarcane juice was in the order of elemental S > ferrous sulfate > gypsum. Aulakh and Dev (1978) found that alfalfa responds to the S in superphosphate better than to that in gypsum, which does not supply P. Swarup and Ghosh (1980) observed that continuous application of SSP increased the sulfate S in soil more than did he treatment with DAP when they were used on an equivalent P content basis.

Badhe and Lande (1980), from a study of S uptake by sorghum and wheat grown on medium black soil (Vertisol) of Maharashtra (India), reported that availability of S, concentration and uptake of S, and dry-matter yields were higher with gypsum than with potassium sulfate.

Sulfur for Soil Reclamation — For the reclamation of sodic soils, gypsum or pyrites are being extensively used in India. Both substances, besides having ameliorating properties for soils, are also good sources of S and are being used for dual purposes. The use efficiency of both substances depends on the fineness of the product and management practices. The results reported by Verma and Abrol (1980) indicate that, when compared on a chemical equivalent basis, gypsum was superior to pyrite on rice and wheat in sodic soils (Table 6.3). Similar conclusions have been drawn by Singh, Hira, and Bajwa (1981) who compared gypsum, pyrites, and elemental S in reclamation of sodic soils. Jaggi (1982) reported that in India 195,000 mt of Amjhore pyrites (a low grade of pyrite) has been used for reclamation of 50,000 ha which produces nearly 200,000 mt additional food.

From column leaching studies in the laboratory Hira and Singh (1980) concluded that the amount of percolating water required for dissolving agriculturalgrade gypsum depends on the particle size of gypsum and exchangeable sodium in the soil. The amount of water required increased from 2.8 to 15.9 cm as the particle size increased from < 0.1 mm to 0.5 mm to 2.8 mm. No more than 4 cm of water was needed to completely dissolve agricultural-grade gypsum of < 0.26 mm particles. Considerable work has also been done on the use of elemental S for soil reclamation as well as a source of the nutrient, but because of its high cost and comparatively low efficiency as a nutrient source, it is not used much in Indian agriculture.

Table 6.3. Comparative effect of gypsum (G) and pyrite (P) on rice and wheat grown on sodic soil in Karnal, Haryana, India.

Treatment (mt/ha)	Grain yield		Sulfur content of grain	
	Rice (kg/ha)	Wheat (kg/ha)	Rice (mg/100 g)	Wheat (mg/100 g)
Control	3 855	19	33.4	171
G1 7.1 mt gypsum	6 707	1 460	45.8	120
G2 14.2 mt gypsum	6 855	3 145	55.3	116
G3 21.3 mt gypsum	7 436	3 600	55.0	109
G4 28.4 mt gypsum	7 239	4 220	56.3	143
P1 3.6 mt pyrite	5 714	146	42.5	169
P2 7.2 mt pyrite	6 038	538	47.0	176
P3 10.8 mt pyrite	6 712	1 350	46.0	129
P4 14.4 mt pyrite	6 914	1 351	52.5	174

Source: Verma and Abrol (1980).

Bangladesh

Deficiencies of S in rice-growing area of Bangladesh have been indicated by a number of studies in the last 20 years (Karim and Majlish, 1958; Karim, Alam, and Rahman, 1970; and Alam and Karim, 1972). Sakai (1980), from an analysis of soils from the upland and lowland rice areas of Bangladesh, concluded that lowland rice soils of Bangladesh were most deficient in S. Many soil analyses and a special symposium on S have indicated the problem of S deficiency in the country. However, there are also contradictory reports. Hoque and Khan (1980) reported that in the autumn season of 1979 the rice cultivar Chandina on farmers' fields did not show any significant effect of S, though the effect of N was most spectacular.

Hoque and Hobbs (1980) reported that in tests conducted by the Bangladesh Rice Research Institute (BRRI) an application of 34 kg sulfate-S as AS increased the yield of rice by 0.1 to 1.3 mt/ha and on farmers' fields by 0.3 to 2.2 mt/ha over and above that attributed to the application of 60 kg N/ha. Best results were obtained with AS as a source of S, and the Aus season rice responded more to S application than did the Aman season transplanted rice. The local varieties responded slightly more to S fertilization than did the new high-yielding varieties.

Frederick (1983) observed that S deficiency is being increasingly recognized as a factor that limits rice production in Bangladesh. In the last 3 years, 51 field trials using gypsum as a source of S indicated that rice yields increased by 0.86 mt/ha. In another set of 25 trials, the increase in yield ranged from 0.22 to 4.2 mt/ha with a mean increase of 1.12 mt/ha. The responses to S were also as dramatic in wheat as in rice. BRRI scientists estimate that about 2.8 million ha of rice suffers from S deficiency.

Sri Lanka

Deficiency of S in tea, coffee, and coconut has been reported from Sri Lanka. De Silva, Anthonypillai, and Mathes (1977) observed that the application of S increased the total fruit yield and the weight of copra. There are not many published reports of responses to S in other crops; however, this may indicate a lack of research rather than the absence of S deficiency in other crops in Sri Lanka.

Indonesia

Blair and Till (1981, 1982) report that S deficiency in Indonesia is very common, particularly in old, highly weathered soils and recently formed soils of volcanic origin. In South Sulawesi, Blair and associates observed a grain yield increase in rice in 18 of 28 sites, and the average response was 19%, ranging up to a maximum of 287% (Blair, Mamaril, and Momuat, 1978, 1979; Blair et al., 1979; and Blair, Momuat, and Mamaril, 1979). Besides increasing grain yield, S also increased the efficiency of N utilization (Table 6.4). Chemical analysis of 254 samples of rice plants collected from Java indicated that 31% were deficient and 42% were marginal in S. Sulfur response has also been observed in upland crops and pastures (Blair, Paulillian, and Samosir, 1978; Blair and Till, 1981).

The S deficiency results not only in lower crop yields but also in lower nutritive quality of grain and forage. In experiments on S-deficient soils in Indonesia, Ismunadji and Zulkarnaini (1978) observed that a change of N source from urea to AS not only increased the yield of rice, but also increased the crude protein and the methionine content in brown rice (Table 4.6). The authors also observed that rice plant samples containing less than

Table 6.4. Rice response to sulfur application and relative efficiency of nitrogen utilization, South Sulawesi, Indonesia.

S application rate (kg/ha)	Grain yield (mt/ha)	Relative N recovery in grain[a]
0	3.33	100
7.5	3.95	114
15.0	4.60	129
30.0	4.93	138
60.0	5.61	155

a. N recovery relative to S_0 treatment.
Source: Blair and Till (1981).

0.1% S indicated an S deficiency; such samples were found in Java (12 locations), Bali (1 location), North Sumatra (3 locations), West Sumatra (6 locations), and South Sulawesi (7 locations). The S content of rice plant tissue ranged from 0.03% to 0.1%.

While studying the effect of source, rate, and time of application of S on flooded rice in Sulawesi, Blair et al. (1979) observed that three sources-AS, gypsum, and elemental S − were equally effective when applied at transplanting time. Elemental S applied 20 days before transplanting was less effective than elemental S applied at transplanting time. This observation is at variance with the observation of Wang, Liem, and Mikkelsen (1976a,b) and Wang (1978) who found elemental S less effective than gypsum for rice production in lowland soils of Amazon Brazil. In another study Ismunadji, Zulkarnaini, and Miyake (1975) reported that sodium sulfate was as effective a source of S as AS in removing S deficiency.

Malaysia

Sulfate-S content of six Malaysian soils ranged from 3 to 155 ppm compared with 7-1 ppm of Iowa soils (Nor, 1981). Responses to S in field crops have been reported from Sarawak province. It has also been reported that marginal deficits of S show up when the deficiencies of other nutrients are removed (Blair and Till, 1981).

Philippines

Lockard, Ballaux, and Liongson (1972) reported responses to S in rice on Luzon soils in two of three experiments. The authors could not find any correlation with the sulfate-S or total fertility of the soil. Kämpfer and Zehler (1967) reported that in field experiments with rice, full potential of the crop was only obtained in the presence of S, and the phosphate uptake by rice was increased in the presence of sulfate.

Thailand

Sulfur deficiencies have been reported in northern Thailand. In one trial, conducted on granite soil over 2 years, a local variety of pigeon pea responded to S and P addition with increases of 75% and 83%, respectively (Andrews and Manajuti, 1980). Keerati-Kasikorn (1982) reports that a great S deficiency is experienced in the highlands and in northeast Thailand but not in the floodplain soils. The total S in Alfisols, Ultisols, and Oxisols was less than 100 ppm. Organic S was a dominant fraction in some northeast soils. Most soils adsorbed sulfate, and the extractable sulfate-S in 68 samples ranged from 1 to 58 ppm. The author also observed that 50 kg S/ha added as gypsum was leached from the soil to 30-cm depth with 225-950 mm rainfall.

A summary of trials by the International Rice Research Institute (IRRI) scientists in Thailand shows that there are widespread S deficiencies in wetland rice. Besides rice, pasture production is also limited by S deficiency in some areas of Thailand. A general level of S application for obtaining optimum production of forages appears to be 30-40 kg S/ha (Blair, 1979).

Papua New Guinea

A 50-year fertilizer experiment was conducted on the southeast coast of Papua New Guinea on coconut, using 1.1 and 2.2 kg muriate of potash (MOP). The responses to K ceased after 1976 when S deficiency became evident, and potassium sulfate had to be substituted for MOP (Sumbak and Best, 1976). In another study on coconut nutrition in Markham Valley of New Guinea, Galasch (1976) reported that the yield of copra increased from 200-500 kg/ha to over 1,700 kg/ha with application of S.

Summary for Asia

From the foregoing it is evident that inherent as well as induced deficiencies of S are widespread in many soils and crops in Asia. Several studies dealing with sulfur in agriculture in Southeast Asian and South Pacific countries have also been summarized in Blair and Till (1983). These studies also indicate occurrences of S deficiency in several of these countries. Crop response to sulfur in Asia and other countries has also been reported in Hoeft (1981). The stronger evidence is available from India, Indonesia, the highlands of Thailand, the lowlands in Bangladesh, and in isolated parts of Malaysia and Sri Lanka. It is possible that, because of lack of published information or inadequate research, S deficiency may actually be even more marked and widespread. Sulfur deficiency affects not only the production of crops but also their nutritive value, which has great significance for food and nutrition of the Asian population.

Though S deficiency in Southeast Asia is very common, there are some areas with an excess of S. Generally these are acid sulfate soils and marshy lands. They cover 3.72 million ha (Van Breemen and Pons, 1978). Of this, nearly 2 million ha is in Indonesia. These soils are generally present in the coastal areas and marshy places where, because of reducing conditions, larger quantities of ferrous sulfide are formed which, on partial oxidation, give rise to sulfuric acid. Various methods of reclamation such as leaching, improved water management, and liming are used for these soils. In such soils, instead of S deficiency, S toxicity is common.

Africa

In Africa the earliest recognition of S as a fertilizer came with the efforts of the European scientists who had major interests in export-oriented crops, including tea, coffee, sugarcane, coconut, groundnuts, and cotton. Serious deficiencies of S were reported from the west and equatorial African countries by the French and British scientists. The S deficiencies and responses to application of S-containing fertilizers in groundnuts and cotton and, to a lesser extent, in cereals are reported from Burkina Faso, Cameroon, Central African Republic, Chad, Benin, Ghana, Ivory Coast, Mali, Niger, Nigeria, Senegal, and Togo.

In east, southern, and central Africa, cotton, groundnuts, alfalfa, legumes, maize, tea, coffee, sugarcane, oil palm, and tomatoes have been reported to respond to S-containing fertilizers. Countries where crop responses to S are reported include Congo, Zaire, Kenya, Malawi, Madagascar, Zimbabwe, Tanzania, Uganda, and Zambia (Beaton and Fox, 1971).

Some of the important reviews on S are Bolle-Jones (1964); Dabin (1972); Martin-Prével (1972); and Richard (1972). The papers presented at the International Symposium on Sulfur in Agriculture held at Versailles in 1972 sum up the situation of the past research in Francophone Africa. The Sulphur 1982 symposium held in London provides additional information, particularly about Nigeria and Kenya (Bromfield et al., 1982). In the following sections we discuss some of the important results for selected countries in west Africa, east Africa, and southern Africa.

West Africa

Richard (1972) has reviewed the S experiments on tropical crops in west Africa. In southern Senegal the responses to P and S are high, whereas in the central region responses to N, P, and K are higher than that to S. However, in both zones a 15%-25% increase in production of groundnuts can be achieved through addition of S.

Twenty-nine field experiments on groundnuts in Senegal showed an average response of 71 kg/ha to S and a total fertilizer response of 522 kg/ha (Bockelee-Morvan and Martin, 1966). There is no doubt that P is the most limiting factor, and S comes next, followed by N and K. Richard (1972) reports that S deficiency in groundnuts becomes visible between 30 and 65 days. Average-to-low responses to S were obtained in tropical ferruginous and ferralitic soils and average-to-high on hydromorphic soils.

Sulfur deficiency in cotton in tropical Africa appeared for the first time with the application of urea N in 1955 in Chad. Experiments on cotton were conducted in tropical Africa by the French scientists from 1965-68; on the basis of their results, Richard (1972) concluded that the S-deficient areas could be grouped in three categories:

1. Areas of high S deficiency   Burkina Faso, Ivory Coast, and Northeast Benin.

2. Areas of medium S deficiency   Cameroon, Central African Republic, and Central Togo.

3. Areas of low S deficiency   Mali, Chad, Southern Togo, Southern Benin, Senegal, and Niger.

Since then, however, the S problem in Africa has increased. Rihard (1972) also concluded that the critical S level in cotton leaf depends upon its P content. Three distinct phases in S response were identified: (I) Crop response to S is always positive even if S is used alone; (II) crop response to S appears only in the presence of other nutrients, particularly N, which induces S deficiency; and (III) no response to S occurs, whatever the condition of fertilization.

The responses to S in tropical soils for various crops in selected African countries are given in Table 6.5. Sulfur deficiency was observed for the first time in French-speaking equatorial Africa in 1955 when S-free urea and nitrates were used in place of AS. Results of more than 200 trials on cotton over a 10-year period are summarized by Braud (1970) in Table 6.6. It may be seen that S deficiency existed in 56% of the trials. Braud (1970) also reported positive S response for cotton in Ivory Coast, based on perennial

Table 6.5. Crop response to sulfur fertilization in selected African countries[a].

Country	Crop	S_0 (kg/ha)	S_1 (kg/ha)	Dose of S applied (kg/ha)	Yield increase per kg S (kg)
Senegal	Groundnuts	1 470	1 558	30	2.9
Upper Volta[c]	Sorghum	1 287	1 464	15	11.8
	Cowpea	688	952	10	26.4
Niger	Groundnuts	1 966	2 078	23	4.9
	Pearl millet	1 734	1 933	29	6.9
Togo	Groundnuts	1 227	1 784	12	46.4
Central African Republic	Rice	2 256	2 305	46	1.0
Benin	Rice	1 707	2 924	36	34.0
	Rice	1 600	2 256	36	18.0
	Rice	2 938	4 476	32	48.0
	Groundnut	1 323	1 967	23	28.0
	Groundnut	2 057	2 535	23	21.0
	Maize	2 630	2 935	46	6.6
	Maize	624	940	36	8.8

a. Basic fertilizers were NPKs.
b. S_0 is the yield of a control (without sulfur). S_1 is the yield with the sulfur treatment indicated.
c. Upper Volta has been renamed Burkina Faso.
Source: Dabin (1972).

Table 6.6. Number of trials grouped by yield of minus sulfur treatment on cotton expressed as % of plus sulfur treatment of NPKs.

Country	Number of trials with			
	Serious S deficiency, <70% yield	Marked S deficiency, 70% – 90% yield	No statistically significant S deficiency, >90% yield	Total trials
Cameroon	1	9	7	17
Ivory Coast − savanna soil	12	16	9	37
− forest soil	–	3	7	10
Benin − northeast	4	3	1	8
− northwest	–	4	3	7
− central	7	6	2	15
− south	2	2	12	16
Upper Volta[a]	7	7	6	20
Mali	–	8	8	16
Central African Republic	1	6	12	19
Chad	–	4	13	17
Togo − north	2	4	4	10
− central	1	4	1	6
− south	–	–	4	4
Total	37	76	89	202

a. Upper Volta has been renamed Burkina Faso.
Source: Braud (1970).

trials where balanced fertilization was provided by NPKS instead of NPK for a few years. It was concluded that, in most cases, S is an indispensable element in the cotton-growing areas of tropical Africa.

It may be observed that all the evidence from French-speaking areas of west Africa is based on field trials on groundnuts and cotton, both of which were export commodities. The evidence of S deficiency in food crops like rice, maize, sorghum, pearl millet, and cowpeas is very scanty. Despite the fact that cotton and groundnuts were regularly fertilized, they suffered from S deficiency. It is conceivable that the cereals and legumes, which were not normally fertilized, would be suffering from incipient or even marked S deficiency. This should become more evident as the traditional agriculture is replaced with modern agriculture.

Because of the serious deficiency of P in this region and the availability of local phosphate rock in most of these countries, ground phosphate rock may be used for direct application, as is done in Mali. This practice can no doubt supply the P requirements, but not the S requirements. Thus, there is a strong case for using partially acidulated phosphate rock (PAPR) in order to improve the possibility of supplying more easily available P and S.

Finally, the paucity of information on available S status of soils in this region seriously limits the formulation of appropriate strategy for fertilizer use.

Nigeria – The first report of S deficiency in west Africa was made by Greenwood in 1948 on groundnuts at Kontagora in northern Nigeria (Greenwood, 1951). The deficiency was also found in Dauro and Kano in Nigeria. Goldsworthy and Heathcote (1963) reported marked responses by groundnuts to S-containing fertilizers. Oke (1967), working in southwest Nigeria, established that extractable S in these soils was low, in many cases below 0.3 ppm, but in some it was as high as 10 ppm. Oke (1969, 1970) reported responses to S by legumes and grasses.

An intensive research project on S was carried out in Nigeria from 1969 to 1974. Bromfield, who led these investigations, concluded the following (Bromfield, 1974d):

1. There is an annual gain of S at a Samaru, Ahmadu Bello University farm, of about 2.8 kg/ha from rain and dust, and a loss of 0.3 kg/ha in drainage, leaving a net S gain of 2.5 kg/ha.
2. A groundnut crop producing 600 kg/ha pods removes 2.5 kg S/ha annually. Thus, at a subsistence level of production the addition of S from the atmosphere and loss by the crop could balance, but in many situations there is evdence of S deficiency.
3. A fertilizer with 1.0:0.7 ratio of P:S will be suitable for groundnuts on these soils.
4. A mixture of ground phosphate rock and S is as good as SSP for groundnuts in Nigeria.
5. Annual application of S is needed for shallow-rooted crops on soils with high S-absorbing capacity.
6. Elemental S oxidizes rapidly and is equivalent to sulfate-S in availability.
7. Use of underacidulated superphosphate was suggested for correcting any serious S deficiency.
8. The apparent recovery of S from the fertilizer sources was 12.1%-25.3% at sites having the higher level of S, and 16.9%-38.7% at sites having the lower level of S.

Kang and Osiname (1976) conducted experiments on the response of maize to S fertilizers from the forest savanna to grassland savanna in Nigeria. Results of these experiments showed that the responsiveness to S fertilization increased as extractable sulfate decreased (Table 6.7). Fox, Kang, and Nangju (1977) observed S responses in cowpeas, but the response varied with the varieties.

Bromfield (1973) reports that of 17 experimental locations with groundnuts as a test crop and phosphate rock as a fertilizer, only one failed to develop S-deficiency symptoms within 5 to 7 weeks after germination. Furthermore, under natural vegetation and unfertilized crops, the soil profile

Table 6.7. Responses of maize to sulfur in Nigerian Savanna soils.

Location	Vegetation zone	Exchangeable SO$_4$-S in Soil[a] (ppm)	Yield response to S fertilizer in Maize (%)
Ikenne	Forest savanna	8.5	+ 7
Ibadan	Forest savanna	3.8	+29
Oyo	Savanna	2.5	+34
Ogbomosho	Savanna	2.8	+98
Ikoyi	Savanna	2.5	+33
Kishi	Savanna	2.8	+15

a. Extractable with monocalcium phosphate solution.
Source: Kang and Osiname (1976).

contained little sulfate; however, S did accumulate for a time after clearing if it was not removed by cropping. Almost all the S applied in 19 years could be accounted for in harvested crops or as residual S in the profile, which showed that erosion and leaching losses had been minimal.

Ghana – Stephens (1960) has summarized the evidence of S deficiency in groundnuts and cereals in both the Voltaian sedimentary and northern groundnut soils, which are generally grouped as ferruginous and ferralitic and receive more than 1,000 mm rainfall annually.

East Africa and Southern Africa

Numerous cases of S deficiencies have been reported from time to time in Tanzania, Kenya, Uganda, Malawi, and Zimbabwe in the east, and in southern Africa under the British rule and Madagascar under French rule. In fact, the classical case of 'tea yellows' attributed to S deficiency was from this part of Africa (Storey and Leach, 1933). Considerable work on correcting S deficiencies through the use of elemental S, AS, gypsum, and SSP has been done in some of these countries.

Sulfur deficiencies in tomatoes (Mehlich, 1970), cotton (Dabin, 1972), sugarcane (Hill, 1963), and wattle (Gosnell, 1964) were reported in Kenya. Of 133 pot trials, 71% showed yield depression of more than 20% without S (Mehlich, 1970). Bolle-Jones (1964) has reviewed the results of past studies on S deficiency in the region most extensively. Bromfield et al. (1982) have reviewed the research work done in Kenya.

Some of the important observations about Tanzania, Uganda, Kenya, Madagascar, Malawi, Zambia, and Zimbabwe are given in the following sections. Though there is little published information about S deficiencies in Ethiopia, the climate and soils, as well as personal discussions with Ethiopian soil scientists, suggest that such deficiencies are likely.

Tanzania — In Tanzania, leaf analysis of finger millet and coffee in the Bukoba area indicated the possibility of S deficiencies. Tea yellows have been reported from the Tukuyu area of Tanzania, which were corrected by S application. Leaf analyses of coconuts and groundnuts in Tanzania have also indicated the possibility that a lack of S may be a serious constraint on these crops (Calton, Vail, and Padhya, 1961). There is some evidence to indicate that S deficiencies now exist for tea near Tukuyu in the southwest and also at the Maruku Coffee Experimental Station in the northwest of Tanzania. Both these soils are ferralitic and receive high rainfall (Bolle-Jones, 1964).

Uganda — Fertilizer experiments with cotton and pastures indicated clearly that there was an S deficiency in the Serere soils, which are representative of a large part of northern and eastern Uganda. These soils are also ferralitic and receive more than 1,000 mm of rainfall annually. Bolle-Jones (1964) further concluded that 50% of the large plateau soils south of Sahara, which are highly weathered ferruginous and ferralitic soils, are deficient in S. Groundnuts and cotton are particularly sensitive to S deficiency; the zone with 600 mm annual rainfall is probably the transition zone, with S deficiency increasing as rainfall increases. Wendt (1970) observed that all pasture species in eastern Uganda responded to P and S, giving yield increases of 40%-100% with an application of 70 kg P_2O_5 and 20-40 kg S.

Kenya — Many areas in the western half of Kenya are S deficient, as indicated by experiments on pastures on sandy loam soils at Kitale and from bioassay work for S in the soils of Soctik and Solai (Mehlich, 1970). The soils of the Songhor region near Lake Nyanza and bottom lands near Machakos as well as the young volcanic soils near Kilimanjaro are also expected to be deficient in S. An extensive investigation on S responses in crops was carried out from 1974 to 1980 by Bromfield and associates. On fertilizer trials at 11 sites, with gypsum and elemental S as sources of S and beans as a test crop, there was marked S responses in 3 sites and in the first harvest, but the response disappeared in the second harvest. Trial sites carrying natural vegetation and old pastures showed marked S deficiencies. The field trials with beans, maize, and groundnuts, which covered a range of soil types and altitudes, showed only occasional responses to S, and all soils were found to adsorb S strongly. It was also observed that the effectiveness of elemental S as a source of S depends on the soil temperature and decreases with the altitude.

The researchers concluded that those soils in which deficiencies would be expected when high-analysis fertilizers containing N and P were used would be the coastal sands and recent alluvial soils, which have both low reserves of S and low S-absorbing capacity. Though these soils are in the high rainfall areas, the accretion of S from atmosphere and the loss of S are balanced at present; with high yields and use of high-analysis fertilizers, however, S deficiencies will appear. Low-cost ground phosphate rock mixed with S in the ratio of 1:0.66 (P:S ratio) was found to be as good as SSP for all soils and crops in this region. For sorbing soils S should be added annually. Bromfield and associates further recommended that when pastures over 3 years old and natural vegetation are broken for cropping, an annual dose application of 15 kg S/ha or 90 kg gypsum should be applied.

Mehlich (1970) has reviewed the crop responses to S in Kenya and reported that in field experiments on sugarcane 42 kg S as gpsum increased the cane yield to 73.7 mt/ha as against 52.5 mt in the control. Likewise, Mehlich (1970) reported that with maize in the central and eastern provinces application of fertilizers without S at planting time reduced vegetative growth in the early stages of development but did not affect the yield, provided the maize was topdressed with S-containing fertilizers. Similarly, topdressing of S in star grass and Rhodes grass produced beneficial effects.

Zimbabwe – Gosnell and Long (1969) have reported S deficiencies in sugarcane in Zimbabwe. According to them, three treatments receiving large amounts of S (gypsum, magnesium sulfate, or elemental S) produced far greater yields of cane and sucrose than did the other treatments. The sucrose production was 25% higher than with the control, which had received N and P but no S.

Grant and Rowell (1976, 1978), while summarizing the results of field experiments on maize in Zimbabwe, concluded that because of the low S status of soils and high degree of weathering of the ferralitic and paraferralitic soils, crops need fertilization with S-containing fertilizer mixtures or gypsum. The experiments on maize at a number of sites in nine districts showed the following:
1. Significant responses to S were obtained on virgin soils or new sites where fertilizers had not been applied in the past, and the degree of response depended on the variety of the crop, site, or soil and increased with liming.
2. Fertilizers containing a minimum of 6.5% S were recommended for supplying enough S along with other major nutrients.
3. The P:S (P_2O_5:S) ratio in fertilizers recommended for Zimbabwe was 2.7:1.0 or 2.1:1.0 instead of 3:1 as indicated by Bixby, Tisdale, and Rucker (1964) in the United States.
4. With an S deficiency, the total S as well as N in leaves was reduced; hence, the N:S ratio did not serve as a good index of S deficiency.
5. The soils that had been heavily fertilized in the past and had built up large amounts of adsorbed sulfate did not respond to S application. This was the case with heavily fertilized soils of tobacco-growing areas of Bindura district.

Vogt (1966) reported results of experiments conducted on corn for 10 areas at the experiment stations and reported marked deficiency and significant response to S (Table 6.8). From these studies it is evident that S as a fertilizer is an important nutrient in agriculture in Zimbabwe and its requirement will grow in the future.

Madagascar — Halais and Girault (1973) reported that the fertilizer used in Madagascar contained no S from 1966 onward. Consequently, an acute S deficiency in sugarcane appeared by 1972. An application of 42 kg S/ha increased the S content of leaf sheath from 0.06% to 0.17%, cane yield from 53 to 77 mt/ha, and sugar percentage from 8.5% to 8.9%. Fertilizers containing N, S, and K were recommended for general adoption on these soils.

Malawi — Extensive S deficiencies, along with P deficiencies, have been observed in the central and northern provinces, in the south Rukuru and Kositu River catchment areas, and in the Phazi area of South Mzimba (Bolle-Jones, 1964). The soils are ferralitic, and most of them receive more than 650 mm of annual rainfall. Jones (1977) reported that the central and northern Malawi soils, because of their sandy nature and highly weathered condition, have low amounts of total and available S and the crops like tea, groundnuts, cotton, and maize show wide variations in their responses to the application of S.

Zambia — In Zambia the soils are acidic, rainfall is high(1,500-2,000 mm), and yields of maize are very low. The soils are deficient in N, S, P, and Fe. Kiyoura (1982) reported that corn yield can be 1.2 mt/ha on farms without fertilizers, 8-9 mt/ha with fertilizers containing S, and 12 mt/ha on experimental stations in the country. Approximately 25-30 kg S/ha is needed to maintain adequate S supply for field crops in Zambia. The Government has placed high priority on attaining self-sufficiency in fertilizer manufacture in the country, and S-containing fertilizers need to be considered as part of the fertilizer supply strategy. It is interesting to observe that highly

Table 6.8. Effect of sulfur-containing fertilizer on the yield of maize in different regions of northern Zimbabwe.

Treatment	Maize yield, kg/ha				
	Mankoya	Mungwi	MPika	Lundazi	MSekera
Control	477	135	712	558	297
S (elemental sulfur)	477	153	981	1 089	359
N (urea or ammonium nitrate)	684	405	972	603	549
NS (urea or ammonium nitrate + S)	981	990	1 647	1 899	892
NS (ammonium sulfate)	1 080	1 287	1 845	2 061	1 053

Source: Vogt (1966).

productive farms are within a radius of 70-150 km from the copper-producing areas; their high productivity can be attributed to the addition of SO_2 from the copper smelting plants.

Congo − Bolle-Jones (1964) reported that the analysis of groundnut leaves indicated an occasional S deficiency, but in general the S status seemed to be satisfactory in the southern part of the country. The soils in the northern part of the country have not been surveyed, but they are poor soils and may be deficient in S.

Summary for Africa

The evidence on crop responses to S in Africa makes it quite clear that, though there is a paucity of published results in scientific journals, S deficiencies are widespread. Field experiments have been conducted on cotton, groundnuts, tea, coffee, maize, millet, rice, oil palm, and pasture grasses by French scientists in Senegal, Burkina Faso, Mali, Ivory Coast, Central African Republic, Benin, Niger, Chad, and Madagascar and by British scientists in Nigeria, Ghana, Kenya, Uganda, Tanzania, Malawi, Zimbabwe, and Zambia. The results provide convincing evidence of S response and inherent S deficiency in many soils of Africa. The S problem may be accentuated in the future with use of S-free fertilizers and more exploitive agriculture. The salient points can be summarized as follows:

1. The results indicate the increasing deficiency of S due to loss of organic matter and native vegetation, soil erosion, leaching, and increasing use of lime.
2. Most of the African soils are highly weathered soils, with coarse texture, low pH, low organic matter, and high phosphate-fixation capacity which accelerates loss of sulfate.
3. Lack of industrialization, the low rate of addition of S from the atmosphere through precipitation, the export of products containing high levels of S, such as groundnuts, palm oil, coconut, tea, coffee, and cocoa, and the import of high-analysis S-free fertilizers are widening the gap between S removals and S additions.
4. Phosphorus is the basic limiting factor in all of this region; S deficiency is no less common, however, and it will be accentuated by the use of more phosphates unless there is a rational use of sulfates.
5. In view of the great demand for food and the lack of development of fertilizer use programs, research on the use of S as a fertilizer becomes most urgent in this region.

Latin America

The possibilities of S deficiency in Latin American soils are great because of the following four factors:

1. Many of the soils are low in organic matter, hence low in reserve of S.
2. There is little SO_2 in the atmosphere because of the low industrial activity.
3. Many of the soils are of volcanic origin or are highly weathered and contain considerable amounts of allophane, kaolinite, and iron and aluminum oxides, which bind sulfates.
4. Leaching possibilities of sulfates are high owing to the high rainfall.

The need for S-containing fertilizers has been demonstrated in Argentina, Bolivia, Brazil, Chile, Colombia, Costa Rica, El Salvador, Guatemala, Mexico, Peru, and Venezuela (Fitts, 1970). Sulfur deficiencies have been observed in maize, sorghum, wheat, cotton, potatoes, bananas, pineapple, coffee, sugarcane, grasses, and forages. Fitts (1970) summarized evidence of S deficiency and responses to S-containing fertilizers in groundnuts, pastures, and cereals in Latin America.

Kamprath (1972, 1981) has reviewed the responses to S in Latin American countries and highlighted the need for S fertilization of crops. At least 30-40 kg S/ha is needed for soils deficient in S. Without application of S, yield will decrease rather quickly on new lands brought into cultivation (Kamprath, 1981). Manuel Arrando et al. (1976) have studied the effect of elemental S on the yield and protein content of wheat.

Brazil

Many experiments on soil fertility and fertilizer requirements for cotton, coffee, soybeans, pasture grasses, and maize have been conducted on Campo Cerrado soils in the State of Sao Paulo, Brazil (De Freitas, McClung, and Lott, 1960). This research has established that S is one of the most limiting factors in crop production on the Campo Cerrado soils in the region from Orlandia to Barretos. McClung et al. (1962), while summarizing the fertilizer experiment data on cotton for 1959/60, concluded that the average yield increase due to S was 53% at four locations and 21% at seven locations (Table 6.9). Mikkelsen, de Freitas, and McClung (1963) reported a mean increase of 25% in corn, and De Freitas, Gomes, and Lott (1972) observed an increase of 55%-82% in coffee (Table 6.9). The response was more marked in the first year after the land was cleared of vegetation, which indicates a very low reserve of S in these soils.

Mikkelson, de Freitas, and McClung (1963), while discussing the importance of S fertilization on Campo Cerrado soils of Brazil, observed that corn response to S was small but significant at Pirasununga and Orlandia, but there was no response at Matao. All three soils had low pH (4.9) and low base saturation and showed high response to Ca, Mg, P, K, and micronutrients. These soils cover nearly 160 million ha of scrub savanna in the central plateau of Brazil. On the basis of research, an application of 30-60 kg/ha of S annually was recommended for these soils. Ammonium sulfate, single superphosphate, and potassium sulfate types of fertilizers were considered preferable for use on coffee plantations. McClung, de Freitas, and Lott (1959) reported that the adsorbed sulfate in Brazilian subsoils, which are highly weathered, become a valuable source of S. The soils containing more than 10 ppm of ammonium acetate-extractable sulfate-S in the B horizon did not respond to application of S, as is evident from pot culture trials in which millet was fertilized wth S (Table 6.10).

Wang, Liem, and Mikkelsen (1976a, 1976b) concluded that S deficiency is a limiting factor for rice production in the lower Amazon Basin in Brazil.

Table 6.9. Effect of sulfur on yields of different crops grown on Brazilian Cerrado soils (based on field experiments).

Crop	Number of locations	S (kg/ha)	Yield[a] (kg/ha)	Yield increase over control (%)	Reference
Cotton	7	0	1 624	–	McClung et al. (1962)
		30	1 971	21	
	4	0	1 377	–	
		30	2 113	53	McClung et al. (1962)
Corn	2	0	4 720	–	
		67	5 909	25	Mikkelsen, de Freitas, and McClung (1963)
Coffee		0	1 344	–	De Freitas, Gomes, and Lott (1972)
		17	2 078	55	
		34	2 384	77	
		67	2 444	82	
		134	2 212	65	

a. Yield is reported as seed cotton for cotton, as grain for corn, and as cleaned coffee beans for coffee.

Table 6.10. Adsorbed sulfate in cropped Brazilian soils and millet response to sulfur fertilization.

Soil	Horizon	Adsorbed sulfate (S ppm)	Yield of dry matter	
			−S (g/pot)	+S (g/pot)
Barau I	AP	2.5	2.5	16.8
	B	12.0	18.0	20.9
Barau 5	AP	4.0	4.3	20.3
	B	21.7	24.5	19.5
Terra Roxa	AP	7.2	12.1	14.8
	B	12.3	19.6	19.2

Source: McClung, de Freitas, and Lott (1959).

The results indicate that when the Varzea marshlands of the lower Amazon are reclaimed for rice production S deficiency develops, which limits rice production especially where high-analysis fertilizers are used. The field trials confirmed that at least 10 kg S/ha is needed for rice production. In the experiments with IRRI varieties of rice and with gypsum as a source of S, the levels of S as high as 1,000 kg/ha did not harm yield. Under field conditions 27 kg S supported two crops, which implies significant positive residual effect. The researchers also concluded that immobilization of S is responsible for reducing the availability of S in residues from previous crops. The same authors also observed that gypsum and AS were equally good sources of S for rice, but elemental S was less effective. Wang (1978) emphasized that fertilizers containing S other than as sulfate are not particularly useful for rice grown in flooded soils.

It was observed that about 20% of the S from one application of 25 kg S/ha was recovered by the rice crop (Wang, Liem, and Mikkelsen, 1976b). The residual effect was studied only on IR22 rice cultivar which was grown on a field receiving 45 kg/ha residual S. It was observed that only 2.8% S was recovered from fresh application of 10 kg S; the rest came from the residual effect of previous applications of 45 kg S.

Venezuela

As reported by the Sulphur Institute (1975), experiments in Venezuela indicated that the grain sorghum crop did not respond to phosphate without S (Figure 6.4).

Costa Rica

According to Fitts (1970), S deficiency in Costa Rica was found responsible for low yield of pineapple, and the yields increased markedly with the addition of S. Valverde, Bornemisza, and Alvarado (1978) observed that, of soil samples from 18 sites, 13 samples gave responses to S and 2 showed no response. For most of the soils sulfate accumulation in the subsoil was noted, which indicates the possibility of subsoil supplying the S needs of the plants at early growth stage. In greenhouse experiments with sorghum, the foliar S content was higher for S-treated soils (1,255 ppm) than for untreated soils (931 ppm). A foliar content of 1,600 ppm S was considered critical for S application to sorghum. The researchers recommend application of S through fertilizer for obtaining good yields of the crop.

Mexico

Mexico, one of the few S-producing countries in the developing world, produces the largest amount of S of any country in Latin America. Infor-

114

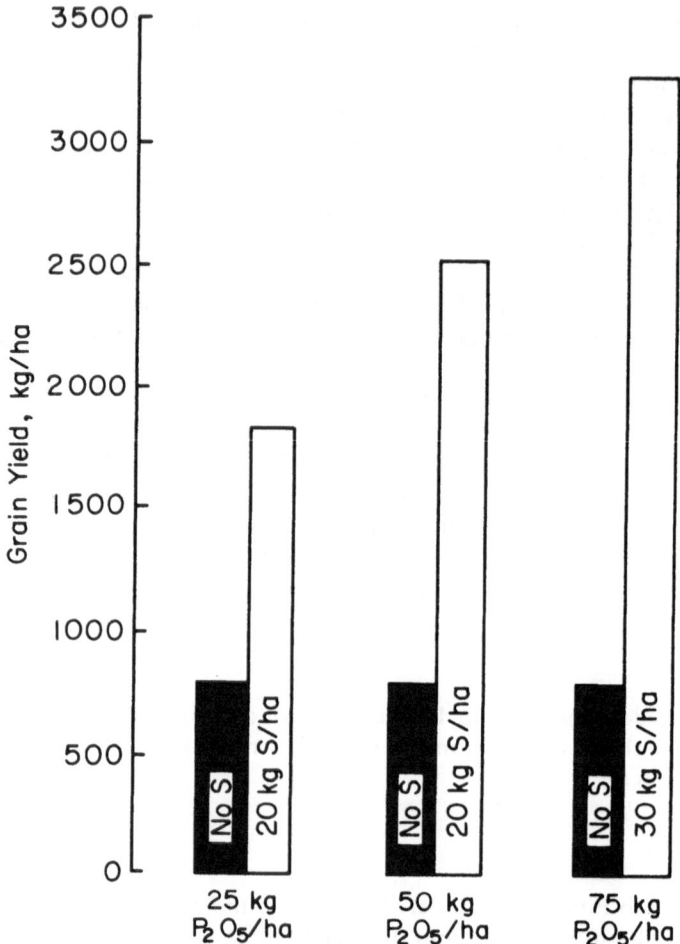

Figure 6.4. Interactions of Sulfur and Phosphates on Grain Sorghum Yield in Venezuela.

mation about S deficiency in soils and crops in Mexico is very scanty, but from recent publications it is evident that there is a growing concern about S deficiency. Amaya (1981), in a recent conference on S in Mexico, drew attention to the S deficiencies in sandy soils and volcanic soils of Mexico. The evidence for the S deficiency comes from the fertilizer experiments on maize, pasture grasses, and oats. The observed increase in grain yield of maize was 5%-14% for sandy soils and 6%-16% for clay soils in response to a shift from S-free to S-containing N and P fertilizers. Huacua and Cajuste (1981), while reviewing the S status andadsorption in Mexican soils, reported that S in soils extractable with phosphate solution varied from 4.8

to 17.9 ppm; the S added to soils was strongly adsorbed. They made a plea for fertilization with S in the soils deficient in this nutrient.

Colombia

Ayala, Guerrero, and Gamboa (1973) reported that in Colombian soils total S varied from 52 to 120 ppm for surface soils and from 26 to 3,020 ppm for subsoils. The mean total concentration of S was 2,027 ppm on the Pasto Highlands (Andepts); 1,766 ppm on the Pacific Plateau (Oxisols); 1,248 ppm on the Tiquerres Highlands (Andepts); 1,204 ppm on the Pulamayo Area (Oxisols Alluvials); and 1,105 ppm on the Ipiales Highlands (Andepts). Andept soils of the warm region had the lowest amount, or 103 ppm of S. The authors have reported 103, 16, 87, 40, and 46 ppm of total, inorganic, organic, exchangeable, and reserve S, respectively, in the surface soils of warm climate regions of the lower altitude.

In field experiments on cassava, Ngongi, Howeler, and MacDonald (1977) observed that in Carimagua and Tranquero, Colombia, when the sulfate-S content of the soil was 4.0-5.0 ppm, potassium sulfate produced significantly higher yields than did the MOP treatment. The MOP + S produced yields equivalent to those of potassium sulfate. Centro Internacional de Agricultura Tropical (CIAT) scientists (CIAT, 1981) conducted field studies in Carimagua to determine the effect of S fertilization on tropical pastures under native savanna conditions; results showed that S is a key element in modifying the soil fertility dynamics as well as the changes in forage availability, protein quality, tanin content, and intake of the forage Desmodium Ovalifolium.

Chile

Schalscha, Estrada, and Galindo (1972) reported that the easily soluble S in soils derived from volcanic ash in southern Chile was rather low, which indicated an S deficiency. However, substantial amounts of organic S and adsorbed S were present in these soils. The average amount of total S ranged from 423 to 1,104 ppm in the surface soil and from 351 to 1,079 ppm in subsoils.

West Indies

Messing (1970) reported that bananas respond to S in the West Indies. Haque and Walmsley (1974) concluded that the S status of the soils in this area was generally low and that all of the soils responded significantly to theapplication of S. Over 90% of the S was in organic form which had a mean C:N:S ratio of 123.4:10:1.08. Sulfate adsorption was significantly related to the percentage of free aluminum oxide in the soils. The authors further concluded that with the increasing use of high-grade compound fertilizers and high-yielding varieties, S problems of the region will increase.

Ecuador

Tergas (1977) reported the results of greenhouse stuies in Ecuador on Latosols treated with elemental S and gypsum to promote growth and nodulation of forage legumes. The authors observed that Centrosema and Dolichos did not respond to treatments with S but soya and sirato did respond.

Nicaragua

Burbano and Blasco (1975) reported that in soils derived from volcanic materials and belonging to the Pacific region of Nicaragua the total S content ranged from 497 to 1,325 ppm and the organic S constituted 7.95%-31.41% of that, which is rather low. Higher concentrations of organic S are found near volcanoes at a depth of 10 cm in the soil.

Summary for Latin America

The available empirical evidence indicates that the status of S in the soils of the Caribbean, Central America, and South America is generally low. Both annual and perennial crops appear to respond to S fertilization. Additional research is needed to identify S-deficient areas and to develop S supply strategies in the context of existing farming systems and fertilizer production facilities.

Residual Effects of Fertilization With Sulfur

There is very little published information on the buildup or depletion of S in soils, particularly in the tropics, with the continued application of S-containing fertilizer under different cropping systems. A few typical cases of long-term experiments that give some indication of the S trends for annual and perennial crops are reported in the following section.

Annual Crops

Nigeria – One such case relates to northern Nigeria, where the experiments were continued for 19 years using cotton, sorghum, and groundnut crop rotation from 1950 to 1961 and continuous cotton thereafter (Bromfield, 1972). The soil is ferruginous in nature, and the clay content increases with depth. It corresponds to an Alfisol according to taxonomy classification.

The soil samples of all plots were analyzed after 19 years of experimentation. Surprisingly, even after 19 years of continuous cropping, 76%-90% of S added as AS or SSP was still present in the soil profile as sulfate-S. These results show that little of the applied S was lost by erosion and leaching. Bromfield (1972) also concluded that an analysis of surface soil may be very misleading for determining S availability to crops because the subsoil S plays an important part in S nutrition. On the other hand, for shallow-rooted crops like groundnuts, it is important to know the available S in the surface soil.

India — The other, more recent case was started in 1971/72 to study the effect of intensive cropping and fertilizer application on a long-term basis on crop yields and soils in India. The experiments are in progress at 10 research stations representing different soils, agroclimatic conditions, and cropping systems. It is too early to interpret the effect of these experiments, but the changes in S status from the experiment at IARI, New Delhi, can be discussed. In a long-term field experiment, involving a multiple cropping system (with pearl millet-wheat-cowpea rotation) and using heavy doses of fertilizers in an alluvial alkaline soil at IARI, New Delhi, it was observed that N and P enhanced the uptake of S; there was also a marked depletion of S in all the plots except the one where S was supplied every year as SSP (Subbarao and Ghosh, 1981). The changes in S after 7 years' cropping are shown in Table 6.11.

Table 6.11. Changes in available sulfur in soil after 7 years of cropping at IARI, New Delhi[a].

Treatment[b]	Initial available S, 1971 (ppm in soil)	Changes in available sulfur	
		1978 (ppm in soil)	Changes over initial value (ppm in soil)
NPK	22.5	7.4	− 15.1
NP	22.5	8.3	− 14.2
N	22.5	8.3	− 14.2
NPK + FYM	22.5	9.2	− 13.3
NPK + S	22.5	23.1	+ 0.6
Control	22.5	10.2	− 12.3

a. Only selected data with comparable doses of N, P, and K are shown in this table. The yearly crop rotation of pearl millet, wheat, and cowpea (fodder) was fixed.
b. NPK dose for wheat and pearl millet was 120 kg/ha N, 25.8 kg/ha P, and 33.2 kg/ha K and sources of nutrients were urea, DAP, and MOP. For cowpeas the dose was 20 kg N/ha, 17.2 kg P/ha and 16.6 kg K/ha. In case of S treatment, SSP was used to supply 120 kg S/ha/year. FYM was applied only to pearl millet @ 15 mt/ha.
Source: Subbarao and Ghosh (1981).

The authors concluded that there was 54.8% to 67.1% depletion of available S in all the cases except where S was applied. After five annual cycles of multiple cropping, Swarup and Ghosh (1980) observed that continuous application of SSP significantly increased the sulfate-S in soil as compared with treatments with DAP. Ghosh (1980) has also reported the changes in crop yields under all the 10 long-term experiments; they indicate increasing S deficiency in some cases.

Perennial Crops

Malawi — Soils of the Mlange district in Malawi are acid, ortho-ferralitic sandy loams or sandy cly loams on which young tea gardens show an acute deficiency of S that disappears after continuous or intermittent use of AS or other S-bearing fertilizers. Grant and Shaxton (1970) reported the results of long-term experiments on tea plantations in Malawi and observed that young plantations require sulfate fertilizers to prevent tea yellows since the virgin soils are inherently deficient in S to a depth of 4 ft. The standard application of AS resulted in a rapid increase in adsorbed sulfate as well as organic S and enriched the soil to a depth of 2 ft in 30 years, but there was little accumulation below 3 ft. The authors also observed that substitution of S-free fertilizers resulted in reduction of S whereas use of AS once in 3 years and perhaps once in 6 years could maintain the sulfate at an adequate level. The effect of various S-containing fertilizers on sulfate-S indicated that more sulfate buildup occurred in treatments with AS.

Brazil — Sulfur deficiency in coffee soils was established as early as 1952 by Malavolta (1952). A number of studies on S fertilization and occurrence of S deficiency in the commercial coffee plantations have been in progress in the past. According to De Freitas, Gomes, and Lott (1972), S fertilization significantly increased the concentration of sulfate-S in leaf samples and also increased coffee yields by 12% in the first biennial, 18% in the second, 117% in the third, and 154% in the fourth biennial. For 10 years of experiments, S application increased yield from 1,340 kg/ha to 2,444 kg/ha.

Conclusions and Research Agenda

1. The results reported in this chapter are a sample of the range of responses to applied S in selected developing countries in the tropics of the world. The results of numerous studies that have not been published are lying buried in local scientific or administrative reports. Some results have been published in local scientific journals. Unfortunately, the need for fertilization with S has not caught the attention of many of the policymakers. Thus, S fertilization for food production has not generally received its due attention.

2. A number of studies have been conducted in greenhouses, and they provide additional information on the problem of S deficiency and the need for S as a fertilizer for crop production in the tropical countries. However, as mentioned earlier, the greenhouse studies have their limitations for assessing S deficiencies and responses in the field. They can only indicate a problem and therefore should be followed with field experimentation.

3. An analysis of the studies discussed in this chapter indicates that deficiencies of S in Asia, Africa, and Latin America are much more serious than is appreciated. The important crops that have been investigated and have shown responses to applied S can be grouped in nine categories, which are summarized in Table 6.12. Most of these crops are importnt food crops, either directly or indirectly through the animal chain system on which most developing countries depend to feed their people. The others are cash crops, some of which are exported to earn foreign exchange for importing food. All of these crops suffer from varying degrees of S deficiency. The problem of food production in these countries is intimately related to balanced fertilizer use, including S as an important component of multinutrient supply strategy.

4. Sulfur deficiency not only affects the crop yield and total production, but it also seriously reduces the nutritive value of the produce, especially the S-containing amino acids. Consequently, S deficiency has serious implications for human nutrition. In cereals, pulses, and oilseeds, S deficiency reduces the S content of S-bearing amino acids. Sulfur deficiency also seriously affects the quality of the forage and protein, and that affects animal production which, in turn, has serious implications for the nutrition of the people. In Latin America and, to a lesser extent, in Africa, where animal meat is an important source of food, S deficiency adversely affects human nutrition. An S deficiency in pastures and forages seriously reduces their yields and affects animal production.

5. All the available information on the subject indicates that S deficiencies exist in the tropics of Asia (10 countries), Africa (23 countries), and Latin America (15 countries). The crops that have shown responses to S and the soils that have been identified as S deficient as well as the countries where they are located are summarized in Figures 6.5, 6.6, and 6.7.

6. The most serious problem is in tropical Asia where, because of the impact of high-yielding varieties, intensive cropping, and high use of inputs including S-free fertilizers, even the latent S deficiency is becoming apparent. In some cases it may be that inherent S deficiencies are becoming acute, in others it may be the induced deficiency. Severely affected crops are oilseeds, groundnuts, pulses and cereals, and other crops. The second region of concern is Latin America. In addition to large areas of chronic S deficiency, many newly cropped areas are experiencing S deficiency, and the problem is becoming compounded by modern agriculture. The more seriously affected crops are soybeans, beans, groundnuts,

Table 6.12. Sulfur deficiency and response research by crops and countries in the tropics: A summary.

Region/country	Crops/crop groups[a]								
	C	R	P	O	S	F	T	A	M
Asia									
Bangladesh	*		*	*		*			
Burma	*								
India	*	*	*	*	*	*	*	*	
Indonesia	*			*			*	*	
Laos	*								
Malaysia				*					
Papua New Guinea				*					*
Philippines	*								
Sri Lanka	*			*			*		
Thailand	*		*						
Africa									
Benin				*		*			
Burkina Faso	*			*		*			
Cameroon	*			*		*			
Chad				*		*			
Congo				*					
Central African Republic				*		*			
Ethiopia							*	*	
Ghana	*			*					
Ivory Coast			*	*		*			
Kenya	*		*	*	*	*	*	*	
Madagascar				*	*				*
Malawi				*			*		
Mali				*		*			
Niger				*					
Nigeria	*	*	*	*				*	
Senegal	*			*		*			
Sudan				*					
Tanzania	*						*		
Togo	*			*		*			
Uganda	*			*		*	*	*	*
Zaire	*								
Zambia	*							*	
Zimbabwe	*			*			*		
Latin America									
Argentina								*	
Brazil	*	*	*	*		*	*	*	
Bolivia	*								
Chile								*	
Colombia	*	*		*				*	
Costa Rica	*								*
Ecuador	*				*				
El Salvador							*		

Table 6.12. Continued.

Region/country	Crops/crop groups[a]								
	C	R	P	O	S	F	T	A	M
Guatemala							*		*
Honduras									*
Mexico	*							*	
Nicaragua									*
Puerto Rico	*				*				*
Trinidad and Tobago	*								*
Venezuela	*					*			

a. Crop groups: C = cereals; R = root crops and tubers; P = pulses; O = oilseeds and oil crops; S = sugar crops; F = fiber crops; T = stimulants; A = forages; and M = miscellaneous and fruit crops such as pineapple and banana.
Star (*) indicates that sulfur research has been conducted. More details and the names of specific crops are availabe in the text.

Figure 6.5. Countries in Which Sulfur Deficiency Exists and Crops on Which Sulfur Responses Have Been Reported in Tropical Asia.

cereals,cotton, pasture legumes, and grasses. In Africa inherent S deficiency is more serious than the induced deficiency, but the latter is becoming serious in newly cropped areas and under modern agriculture based on high-yielding varieties. In all three regions coffee and tea are well recognized S-responsive crops, and they need regular applications of S, in accordance with the S status of soils and the desired production levels.

7. The S problem is more complex and difficult than those of the conventional N, P, and K fertilizers for the following reasons:

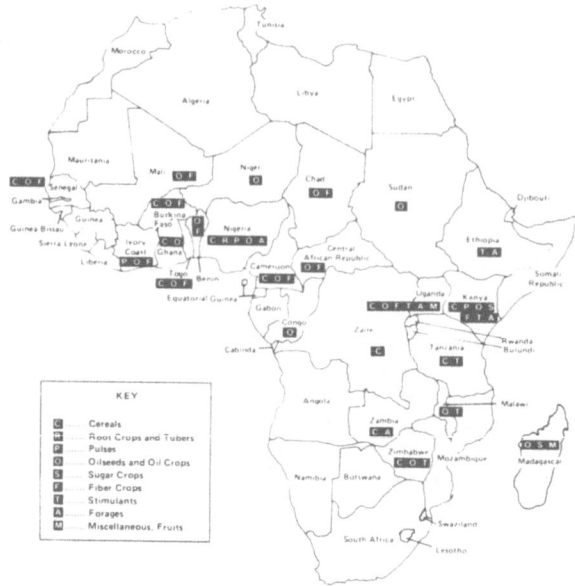

Figure 6.6. Countries in Which Sulfur Deficiency Exists and Crops on Which Sulfur Responses Have Been Reported in Tropical Africa.

a. The use of both lime and phosphates accelerates S loss from soils; thus, the conventional soil fertility management programs of tropical soils are accentuating S deficiency.

b. Unlike phosphates, S, whether added in sulfate or elemental S form, is more liable to serious leaching. The S in the organic matter of the soil is being depleted and lost rapidly in the tropical countries.

c. Most of the fertilizer response research in the developing countries is being carried out on NPK products, and S research is not receiving much attention.

d. The results of agronomic reserch on S from the developed countries or temperate zones have very little applicability in the tropical soils and crops because of the differences in the nature of soils, cropping systems, and environments.

e. In some countries the only sources of S for agriculture are gypsum, pyrites, phosphogypsum, and byproducts of the fertilizer, chemical, and agricultural industries. The amounts available, the crop response, the potentialities for use, and the technology of use for these products have yet to be seriously assessed.

f. There is little information on the recovery of S by the crop and its recycling into the soil fertility management system. Research on improving efficiency has not received any attention.

Figure 6.7. Countries in Which Sulfur Deficiency Exists and Crops on Which Sulfur Responses Have Been Reported in Tropical Latin America.

g. Many of the developing countries export commodities that are great users of S. Although serious efforts are being made to correct the imbalances in use of N, P, and K in fertilizers, little attention is given to the growing imbalance of S.

h. The areas that have a serious inherent deficiency of S or that are likely to become deficient have not been adequately delineated. Coordinated international and national effort is necessary to stimulate research on S as a fertilizer nutrient.

7 Estimating Sulfur Requirements, Supplies, and Gaps

The empirical evidence presented earlier indicates that S deficiency has become a problem in most tropical countries. Unless appropriate corrective actions are taken, S deficiency poses a threat to national and international efforts to accelerate food and agricultural production in these countries. However, in order to design nationally acceptable and economically viable S supply strategies, there is a need to accurately estimate S requirements, supplies, and gaps. The primary purpose of this chapter is threefold: (1) to estimate S requirements for crop production and fertilizer manufacturing, (2) to determine trends in S supply from fertilizer use, and (3) to estimate S gaps between requirements and supplies. The empirical estimation of the mathematical model deals with major crops, selected countries and world regions, and S-containing fertilizers from 1960 to 2000.

Determinants of Sulfur Uptake

The amount of S removed from soil or taken up by field crops depends on several factors, some of which are still not known. At least four of these factors exert major influence on the amount of S removed by field crops; they are (1) crop and the cropping pattern, (2) average crop yield, (3) the area under each crop, and (4) cropping intensity.

A generalized relationship between crop yield and fertilizer S uptake is developed in Figure 7.1. Under a subsistence cropping system using traditional technology, the amount of S uptake at low crop yields is rather small. Most of this need can be met by the existing soil-crop-environment system. However, S uptake increases as crop yields increase. The additional S uptake must be supplied by supplemental S sources. Beyond that required for maximum crop yield, additional S application will not add to grain yield, mainly because of a relatively fixed relationship between N and S in protein synthesis.

Sulfur uptake for most countries is gradually increasing because (1) crop yields are increasing, (2) the cropping pattern is changing, (3) the area under different crops is expanding, and (4) crop cultivation is becoming more intensive because of an increase in cropping intensity.[1] However, the relative magnitude of these changes varies from one country or region to another.

---

1. Cropping intensity = [cropped area/cultivated area] 100. The cropping intensity is an index of the intensity of land use and is expressed in percentageterms on an annual basis. In this context it accounts for multiple cropping or fallow.

126

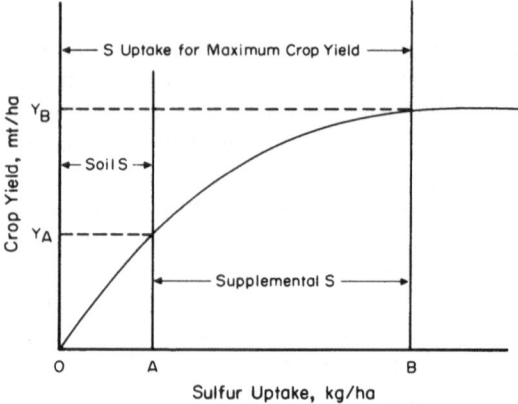

Figure 7.1. A Generalized Relationship Between Crop Response and Sulfur Uptake.

Determinants of Sulfur Replacement Requirements

The amount of S needed to replace that taken up by the crops is generally more than the amount removed from the soil, regardless of the S supply source. The S replacement requirements are determined primarily by the efficiency with which the crops use the S. The S use efficiency in turn depends on (1) crop and crop variety, (2) soil, (3) irrigation regime, (4) environment, and (5) S supply source. Various S supply sources include (1) soil and crop residue, (2) atmosphere, (3) irrigation water, and (4) chemical fertilizer and soil amendments containing S. The amount of S supplied through soil, atmosphere, and irrigation water is generally beyond the control of individual farmers.

Relatively little is known about S use efficiency, residual effects, and loss mechanisms of applied fertilizer S, especially for crops under field conditions in the tropics. According to Noggle (1980) and Beaton et al. (1974), S use efficiency and loss mechanisms, to some extent, are similar to those with N. Beaton et al. (1974) have proposed a replacement factor of 1.75 which implies that S use efficiency is approximately 57%. Kiyoura (1982), on the other hand, has indicated that only about 20% of the S applied by S-containing fertilizers is absorbed effectively by crops, which implies a replacement factor of 5. The available evidence from tropical countries also indicates that S use efficiency, particularly in sandy soils, is about 20%-30%, which implies a replacement factor between 3 and 5.

Model for Estimating Sulfur Requirements

A simple mathematical model used to quantify S requirements is developed below. Let

1. $U_{ijt} = r_i P_{ijt}$,

2. $P_{ijt} = A_{ijt} * Y_{ijt}$, and

3. $U_{jt} = \Sigma_i U_{ijt}$.

Where

U_{ijt} → Sulfur uptake (mt) for ith crop/crop group (out of several selected crops/crop groups) in jth country/region (out of several selected countries/regions) during time t;

r_i → Sulfur uptake coefficient for ith crop;

P_{ijt} → Total production (mt) of ith crop/crop group in jth country/region during time t;

A_{ijt} → Total harvest area (ha) under ith crop/crop group in jth country/region during time t;

Y_{ijt} → Average yield (mt/ha) for ith crop/crop group in jth country/region during time t; and

U_{jt} → Total S uptake (mt) for jth country/region during time t.

Since S replacement requirements are generally more than uptake, an allowance has been made to account for S use efficiency as follows:

4. $R_{ijt} = \beta_k U_{ijt}$, and

5. $R_{jt} = \Sigma_i R_{ijt}$.

Where

R_{ijt} → Sulfur replacement requirements (mt) for ith crop/crop group in jth country/region during time t;

β_k → Sulfur replacement coefficient for kth S use efficiency regime; and

R_{jt} → Total S replacement requirements (mt) for jth country/region during time t.

Finally, the relative share (%) of individual crops/crop groups in total S requirements is estimated as follows:

6. $S_{ijt}^U = [U_{ijt}/U_{jt}] \, 100$, and

7. $S_{ijt}^R = [R_{ijt}/R_{jt}] \, 100$.

Where

S_{ijt}^{U} → Percentage share of ith crop/crop group in total S uptake for jth country/region during time t; and

S_{ijt}^{R} → Percentage share of ith crop/crop group in total S replacement requirements for jth country/region during time t.

The list of crops/crop groups, the list of countries/regions, the S uptake coefficients, S replacement coefficients, and other information used in estimating S requirements are developed in the subsequent sections.

Data for Estimating Sulfur Requirements

Most of the economic and technical data used in this study to estimate S requirements have been obtained from secondary sources. The study deals with the following set of countries and regions (referred to as j in the model):

1. India
2. Indonesia
3. Philippines

4. Kenya
5. Niger
6. Nigeria
7. Sudan
8. Zimbabwe

9. Brazil
10. Colombia
11. Mexico

12. Far East
13. Africa
14. Latin America

15. World

For each of these countries/regions, S requirements are estimated for 9 separate years with 5-year intervals in between. The estimates are based on (1) actual data for 1960, 1965, 1970, 1975, 1980 and (2) projections for 1985, 1990, 1995, 2000. The results for each year refer to 3-year simple averages centered on years shown in order to avoid weather-related variations in crop production. For example, crop production in 1980 refers to a 3-year average of production in 1979, 1980, and 1981.

Although forage crops, vegetable crops, and fruit trees were not included,

25 of the more important crops were used in making S requirement estimates. This does not mean that each country grows all of these crops. Furthermore, the relative importance of different crops varies from one country to another. These crops were then grouped into different crop groups depending upon their nature. The list of all 25 individual crops, 7 broad crop groups and 15 specific crop groups (referred to as i in the model) is given in Table 7.1. In most developing tropical countries, almost all of the chemical fertilizer consumption is generally accounted for by these 25 crops.

Three sets of data form the core of the information needed to estimate the model. These are (1) crop production, (2) S uptake coefficient, and (3) S replacement coefficient. The crop production data for individual crops were obtained from FAO (1982) for 1959 through 1981. The data were further rationalized from other relevant sources. The production data from 1981 to 2001 for specific crop groups were based on projections. These projections were made by using (1) average annual growth rate in crop production and (2) crop production in the initial period. The initial condition refers to production during 1980, which was a 3-year average centered on 1980. The average production growth rate was calculated by estimating,

Table 7.1. List of field crops and crop groups used in estimating sulfur requirements.

Broad crop groups	Specific crop groups	Individual crops
1. Cereals	1. Wheat	1. Wheat
	2. Rice	2. Rice
	3. Maize	3. Maize
	4. Millet	4. Millet
	5. Sorghum	5. Sorghum
2. Pulses	6. Pulses and legumes	6. Beans
		7. Broad beans
		8. Peas
		9. Chick peans
		10. Lentils
3. Oil crops	7. Oilseeds	11. Rapeseed
	8. Soybeans	12. Linseed
	9. Groundnuts	13. Sunflower seed
	10. Oil palm	14. Safflower seed
		15. Sesame seed
		16. Soybeans
		17. Groundnuts
		18. Oil palm
4. Roots and tubers	11. Roots and tubers	19. Potatoes
		20. Sweet potatoes
		21. Cassava
5. Sugarcane	12. Sugarcane	22. Sugarcane
6. Cotton	13. Cotton	23. Cotton
7. Stimulants	14. Coffee	24. Coffee
	15. Tobacco	25. Tobacco

using the least-squares method, a regression equation of the following logarithmic form:

8. $\text{Log } P_{ijt} = a + bt + e_t.$

Where

P_{ijt} → Production of ith crop/crop group in jth country/region during year t.

a → Intercept

b → Regression coefficient

t → Time

e_t → Error term

The estimated value of b ($b = \log [1+g]$) is used to estimate g ($g = [\text{antilog } b] - 1$) which is the least-squares estimate of average annual growth rate. The advantage of this approach is that all the relevant time-series data on crop production are used to estimate the average growth rate. This equation is estimated by using time-series data for 12 years from 1970 to 1981. The estimated average annual growth rates were further rationalized in view of (1) growth rates used by various international organizations (including FAO, IFPRI, and World Bank) in their projections, (2) national government policy with respect to production of a particular crop, (3) land and other resource constraints, (4) yield expansion potential, and (5) national and international agricultural research policy. Any negative growth rate was equated to zero, and any growth rate above 5.0 was equated to 5. These assumptions were needed to avoid any unrealistic production trends in the future.

The S uptake coefficients (r) used in the study are reported in Table 7.2. Most of these estimates were derived from Malavolta (1979). However, some of these estimates for S uptake by specific crop groups were modified in view of results from other similar studies. Finally, because of lack of appropriate information, two levels of S replacement coefficient (β) were used in determining S replacement requirements. These were (1) 1.75, which implied 57.14% apparent S use efficiency and (2) 3.50, which implied 28.57% apparent S use efficiency. The actual S use efficiency, and hence S replacement requirements, may fall within this range.

Estimated Aggregate Sulfur Requirements

Sulfur requirements are classified into three categories. These categories include (1) S uptake, .e. amount of S actually taken up by field crops; (2) S replacement requirements, i.e., amount of S needed to replace the removed

Table 7.2. Average sulfur uptake and sulfur uptake coefficients for specific crop groups.

Specific crop groups	Average sulfur uptake[a] (kg/mt)	Sulfur uptake coefficient (r)
1. Wheat	4.0	0.0040
2. Rice paddy	3.0	0.0030
3. Maize (corn)	4.0	0.0040
4. Sorghum	5.0	0.0050
5. Millet	8.0	0.0080
6. Pulses and legumes	8.0	0.0080
7. Roots and tubers	0.5	0.0005
8. Oilseeds	12.0	0.0120
9. Cotton (seed cotton)	15.0	0.0150
10. Groundnuts (with shells)	6.0	0.0060
11. Sugarcane (cane)	0.3	0.0003
12. Tobacco (dry leaves)	6.0	0.0060
13. Coffee beans	13.0	0.0130
14. Soybeans	8.0	0.0080
15. Oil palm (nuts)	1.5	0.0015

a. Calculated by dividing total sulfur uptake by average crop yield.

S, irrespective of S supply source; and (3) S requirements by the fertilizer industry in the form of S or sulfuric acid needed to produce S-containing fertilizers and phosphoric acid.

The estimated aggregate S uptake from 1960 to 2000 is reported in Table 7.3 for selected tropical countries, developing tropical regions, and the world. The estimated uptake from 1960 to 1980 is based on actual data, whereas that from 1985 to 2000 is based on projections for crop production. The highest S uptake is in India for the Far East region, and in Brazil for the Latin America region. In addition to having the large agricultural sector, these countries are also experiencing rapid technological change and shifts in established cropping patterns. These changes are reflected in regional S uptake. The individual countries in Africa as well as the African region appear to be rather static in comparison with other countries and regions as far as S uptake is concerned. There are several reasons for this situation, including subsistence agriculture, static and in some cases even declining crop yields, and shift in the cropping pattern in favor of those crops that have relatively low S requirements.

The S replacement requirements for the selected tropical countries and regions, and for the world as a whole, are reported in Table 7.4. The replacement requirements are estimated for two scenarios. In scenario I, the S replacement coefficient is 1.75 (implied use efficiency for applied S about 57%), which may be the case in temperate climatic conditions. In scenario II, the S replacement coefficient is 3.50 (which would double the S replace-

Table 7.3. Estimated aggregate sulfur uptake by field crops in selected developing countries and world regions, 1960 – 2000. ('000 mt of S)

	1960	1965	1970	1975	1980	1985	1990	1995	2000
Asia									
India	509	524	651	688	784	886	1 008	1 155	1 333
Indonesia	64	70	87	102	130	159	195	239	294
Philippines	21	24	30	38	45	56	70	87	108
Africa									
Kenya	9	9	13	15	15	16	17	19	21
Niger	6	8	11	9	13	16	19	24	29
Nigeria	56	62	66	59	66	72	80	88	98
Sudan	29	18	28	31	30	36	44	53	64
Zimbabwe	7	8	12	15	16	19	22	25	29
Latin America									
Brazil	157	171	205	284	349	412	490	587	708
Colombia	21	22	26	32	38	46	57	70	86
Mexico	64	94	105	110	137	156	179	206	240
Far East	776	859	1 053	1 140	1 340	1 549	1 799	2 100	2 463
Africa	224	271	314	325	327	346	365	387	411
Latin America	404	488	561	687	844	984	1 153	1 356	1 603
World	4 582	5 214	5 997	6 729	7 911	9 117	10 542	12 231	14 238

ment requirements), which may be the case in tropical conditions, especially for sandy soils. Even in scenario I, the more efficient system, worldwide S replacement requirements have increased from 8 million mt in 1960 to 14 million mt in 1980, and they are expected to increase to 25 million mt in the year 2000. Except in African countries, S replacement requirements are expected to increase rapidly in most tropical countries and regions. These quantitative estimates of S replacement requirements form a basis for designing an effective and economical S supply strategy.

In addition to being a plant nutrient, S is used in the fertilizer industry to manufacture (1) S-containing fertilizers such as AS and SSP and (2) wet-process phosphoric acid used to produce TSP and ammonium phosphates. Although it may be desirable to make estimates of the total S requirements for each country, in the fertilizer industry the lack of accurate data precludes this analysis. In particular, most countries are involved in fertilizer trade, either as exporters or as importers of certain types of fertilizer materials. The S requirements of the fertilizer industry are created at the points where the fertilizer materials are produced, rather than where they are consumed. The estimated S requirements for the world S industry are reported in Table 7.5. These estimates include S needed to manufacture AS, SSP, TSP, and ammonium phosphates.

Table 7.4. Estimated aggregate sulfur replacement requirements by field crops in selected developing countries and regions under alternative sulfur replacement coefficient scenarios, 1960 – 2000.

	Scenario I, S replacement coefficient 1.75[a]. (' 000 mt of S)				
	1960	1970	1980	1990	2000
Asia					
India	891	1 139	1 372	1 764	2 332
Indonesia	112	152	227	341	514
Philippines	38	53	80	122	190
Africa					
Kenya	15	23	26	30	37
Niger	11	19	22	34	51
Nigeria	99	115	116	139	171
Sudan	51	49	53	76	113
Zimbabwe	12	21	29	38	51
Latin America					
Brazil	275	359	611	858	1 239
Colombia	36	45	67	99	150
Mexico	113	184	239	313	419
Far East	1 358	1 843	2 345	3 149	4 310
Africa	392	549	573	639	720
Latin America	707	982	1 477	2 017	2 805
World	8 018	10 494	13 844	18 448	24 917

a. Implied S use efficiency 57.1%. In Scenario II (S use efficiency 28.6%), the estimated S replacement requirements will be double those of Scenario I.

These estimates are based on fertilizer production needed to meet consumption requirements, as projected by FAO/UNIDO/World Bank. The S requirements for the world fertilizer industry are estimated to increase from 26.5 million mt in 1980/81 to 41.1 million mt in 1990/91 and to 64.7 million mt in 2000/01. These requirements are in addition to S replacement requirements for agricultural crops. However, S contained in AS and SSP will be used to satisfy part of the replacement requirements. Even at the least favorable scenario (S replacement coefficient : 3.50), the projected S requirements in the fertilizer industry are more than the projected S replacement requirements for agricultural crops in the corresponding years. Clearly there is need to make economic use of S which is lost in discarded phosphogypsum in order to meet part of the S replacement requirements.

Table 7.5. Estimated sulfur requirements for world fertilizer industry[a]. ('000 mt)

Year	Consumption for[b]		Sulfur requirements for			
	N	P_2O_5	AS[c]	SSP[d]	TSP/AP[d]	Total
1980/81	60 300	31 500	3 300	3 200	20 000	26 500
1985/86	73 100	38 500	4 300	3 900	24 400	32 600
1990/91	86 000	48 900	5 000	5 000	31 100	41 100
2000/01	145 500	76 200	8 500	7 800	48 400	64 700

a. Based on fertilizer production needed to meet projected fertilizer consumption requirements.
b. Actual consumption for 1980/81 is taken from FAO (1983). Projections for 1985/86 and
   1990/91 are from FAO/UNIDO/World Bank (1983); for 2000/01, they are from UNIDO
   (1978). Projections for 1985/86 are the most recent and hence reflect the current economic
   trends. The projections for 2000/01 are the oldest and appear on the high side.
c. Sulfur contained in AS. Estimates for 1980/81 are based on actual AS consumption, containing 21% N and 24% S. During 1980/81, AS accounted for 5.9% of N production and
   4.7% N consumption. In the following years, AS is assumed to account for 5% of projected N consumption.
d. Based on a 3-year average for P_2O_5 capacity and production, the contribution of individual materials was approximately 15% by SSP, 25% by TSP, 50% by ammonium
   phosphate (AP), and 10% by nitrophosphates. These shares are assumed to prevail in the
   following years. Sulfur consumption is assumed to be 0.68 mt/mt of P_2O_5 from SSP;
   0.70 mt/mt of P_2O_5 from TSP; 0.92 mt/mt of P_2O_5 from AP (MAP and DAP); and zero
   for nitrophosphates.

Estimated Sulfur Requirements by Crops

In order to design economically viable S research programs and develop S
supply and S pricing strategies, it is important to determine S uptake by
crops.

The estimated proportions of S uptake by specific crop groups are reported in Figure 7.2 for four major tropical countries and in Table 7.6 for three
tropical regions and the world. The results reported in the table and diagram are designed to indicate (1) the percentage share of individual crops
or crop groups in aggregate S uptake in a particular year and (2) shift in
percentage share for individual crops or crop groups over time from 1960
to 1980 to 2000. The percentage share of individual crops or crop groups
in aggregate replacement requirements may not be the same as in uptake
since it depends on their relative S use efficiency.

In each of the countries and regions, S uptake is dominated by cereals.
The relative share of cereals during 1980 varied from around 33% in Brazil
to 81% in Nigeria. For the world as a whole, cereals accounted for 66% during 1980. The relative share of cereals is expected to shift in the year 2000,
resulting in a decline in Brazil and an increase in Nigeria and the world as
a whole. In Brazil the relative share of oil crops (primarily soybeans) in-

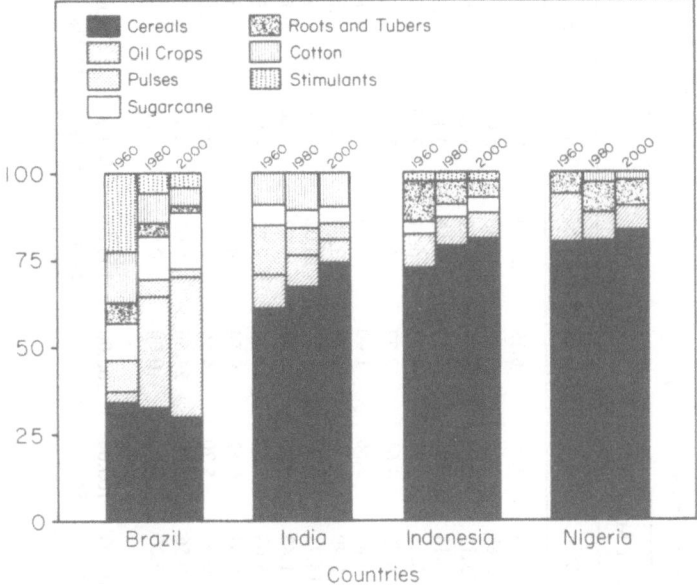

Figure 7.2. Estimated Proportion of Sulfur Uptake by Broad Crop Groups in Selected Developing Countries.

creased from 4% in 1960 to 32% in 1980 and is expected to increase to 41% in the year 2000. Furthermore, within cereals the share of individual crops varies by country and region and over time in a particular country and region. For example, the percentage share of rice during 198 was 29% in India and almost 68% in Indonesia. The relative share of wheat in India increased from 8% in 1960 to 18% in 1980 and is expected to be 27% in the year 2000. The share of pulses is declining, partly because there has been relatively little change in their production. Other more profitable crops are replacing pulses in the cropping pattern. Similarly, for the world as a whole, the relative share of millet, pulses, root crops, cotton, groundnuts, and coffee is declining, whereas that for maize, oilseeds, and soybeans is increasing.

One must be careful in interpreting these shares since they may change in the future depending upon (1) the level and impact of technological changes for different crops, (2) relative price ratios for different crops, and (3) government policy with respect to these individual crops or crop groups.

Estimating Sulfur Requirements From Nitrogen and Phosphorus Use

An alternative approach to the estimation of S requirements for crop production relates S needs to applied N and P_2O_5 from chemical fertilizers.

Table 7.6. Estimated proportion of sulfur uptake by specific crop groups in developing world regions. (%)

Crop/crop group	Far East			Africa			Laten America			World		
	1960	1980	2000	1960	1980	2000	1960	1980	2000	1960	1980	2000
Wheat	7.52	13.96	19.92	5.22	5.65	4.50	8.84	7.07	6.19	21.04	22.43	22.97
Rice	40.35	40.93	41.75	4.10	5.49	6.88	5.88	5.51	5.93	15.48	15.04	14.15
Maize	4.64	5.55	5.03	12.57	17.21	16.84	23.13	22.03	18.43	18.37	21.28	24.82
Sorghum	5.34	4.65	6.32	15.42	11.52	10.83	2.13	7.36	10.30	3.56	4.12	3.43
Millet	8.61	5.93	3.23	27.47	21.97	23.49	0.63	0.24	0.33	6.31	2.89	1.70
Pulses	10.59	5.66	3.09	4.87	5.48	4.36	6.62	4.61	2.77	5.06	3.77	2.09
Root crops	1.60	2.11	3.04	6.48	8.30	10.18	3.86	2.66	1.40	5.10	3.15	1.88
Oilseeds	3.83	3.35	2.18	1.20	1.73	1.69	5.49	4.46	5.06	3.80	4.65	4.81
Cotton	7.03	7.26	5.58	6.27	6.55	5.21	14.73	9.04	6.20	10.02	8.21	6.28
Groundnuts	4.18	3.40	2.44	9.24	7.00	5.57	1.43	0.80	0.42	1.77	1.44	0.88
Sugarcane	4.88	5.35	5.06	1.73	3.11	4.72	15.08	12.87	13.26	2.73	2.87	2.90
Tobacco	0.51	0.43	0.30	0.47	0.44	0.74	0.64	0.52	0.54	0.47	0.41	0.31
Coffee	0.31	0.52	0.75	4.33	4.79	3.82	10.92	5.20	4.56	1.23	0.85	0.69
Soybeans	0.62	0.84	1.21	0.07	0.43	0.90	0.56	17.56	24.56	5.02	8.86	13.07
Oil palm	0.01	0.07	0.11	0.55	0.33	0.26	0.06	0.06	0.06	0.03	0.03	0.04
Total %ᵃ	100.00	100.00	100.00	100.00	100.00	100.00	100.00	100.00	100.00	100.00	100.00	100.00
('000 mt S)	(776)	(1 340)	(2 463)	(224)	(327)	(411)	(404)	(844)	(1 603)	(4 582)	(7 911)	(14 238)

a. Totals are approximate due to rounding of data.

This approach is based on the agronomic results that indicate a rather fixed relationship between uptake of N and S, and uptake of P_2O_5 and S. According to the available literature, the ratio between N and S uptakes for the maximum amount of protein synthesis is about 15:1, and between P and S it is about 1:1 (or 2.3:1 when P is expressed as P_2O_5) (Beaton et al., 1974; Stewart, 1969; Stewart and Porter, 1969).

On the basis of agronomic studies and after making appropriate adjustments for nonfertilizer supply sources and nutrient losses, Beaton et al. (1974) proposed that maintaining a 5:1 N:S ratio and a 3:1 P_2O_5:S ratio in applied fertilizer will provide a balanced supply of N, P_2O_5 and S. Sulfur needs for plant nutrient purposes can therefore be estimated by using these ratios (N:S = 5, P_2O_5:S = 3) in conjunction with N or P_2O_5 needed or actually applied. This method of estimating S needs has also been endorsed by others, including British Sulphur Corporation (1983a), Hignett and Stangel (1982), Meyer (1977), and Tisdale and Platou (1981). This method may provide adequate plant nutrient S to attain the maximum efficiency of applied N and P_2O_5.

This approach appears to be appropriate for estimating the S needs of individual crops under temperate climatic conditions. Even under temperate conditions, however, there is a need for more extensive experimentation under farmers' field conditions in order to establish a realistic range of N:S and P_2O_5:S ratios to be maintained in fertilizer applications. As far as tropical countries are concerned, such an approach to estimating S needs could be rather misleading. Some of the factors that need to be considered in establishing these ratios include (1) type of soil and S supply in the organic matter in thesoil, (2) S supply from the atmosphere and in precipitation, (3) S supply from irrigation water, (4) S supply through crop residues and farmyard manure, (5) magnitude and mechanisms of S losses, (6) cropping pattern, (7) level of crop yields, and (8) sources and amount of the N, P_2O_5, and S supplies. These factors, which influence S supply and requirements, vary not only from tropical countries to temperate countries but also from one country to another. Furthermore, as has been suggested by Stewart and Porter (1969), even the total N:total S ratio varies from 4 to 55 for different crops.

The appropriateness of these ratios in determining adequate S needs is examined by using N, P_2O_5 and S consumption data from India and Brazil. The results are reported in Table 7.7. According to estimated N:S ratios, the S supply in Brazil has always been more than adequate, whereas in India S supply was more than adequate until the mid-1970s. On the other hand, according to the P_2O_5:S ratio the S supply was adequate until the mid-1970s in Brazil and India. At present, the agronomic results indicate S deficiencies in both India and Brazil.

Table 7.7. Estimated nitrogen:sulfur and phosphorus:sulfur ratios in aggregate fertilizer consumption in India and Brazil.

Year	N:S ratio		P_2O_5:S ratio	
	India	Brazil	India	Brazil
1960/61	0.9	0.9	0.2	1.1
1965/66	1.4	0.7	0.3	0.9
1970/71	3.8	1.1	0.9	1.5
1975/76	8.4	1.5	1.6	3.8
1980/81	14.1	1.8	4.3	4.1

Trends in Fertilizer Sulfur Supply

The purpose of this section is to estimate S supply from S-containing fertilizer materials used in tropical countries. Historically, the general trend in fertilizer production and use has been away from S-containing to S-free nitrogen and phosphate fertilizers (IFDC, 1979).

In the latter part of the nineteenth century and early part of the twentieth century, major sources of N included Chilean nitrate, calcium cyanamide, and AS. Ammonium sulfate, which contains 21% N and 24% S, was a rather important source of N until the 1960s not only because of its agronomic effectiveness but also because of the varied sources of its production; it is a byproduct from coke ovens, a coproduct from caprolactam, and a direct product from the reaction of sulfuric acid with synthetic ammonia. In the 1960s, however, ammonium nitrate (33% N and no S) became an important source of N. Finally, in the 1970s urea (46% N and no S) became a dominant source of N worldwide, and particularly in developing countries. In some developed countries direct application of ammonia and liquid compound fertilizers replaced the use of most straight fertilizers.

As discussed by Jacob (1964) and Slack and Hardesty (1964), SSP (18% P_2O_5 and 12% S) was first produced commercially in 1842 by Lawes in England. Among commercial phosphate fertilizers, SSP was the most important source of P_2O_5 in the latter half of the nineteenth century and the first half of the twentieth century. Even though phosphoric acid was commercially produced in the 1870s, SSP provided over 60% of the world's phosphate as late as 1955. Triple superphosphate (46% P_2O_5 and 1% S) became an important phosphate fertilizer in the 1950s. Ammonium phosphates (MAP with 55% P_2O_5, 11% N, and 1% S; and DAP with 46% P_2O_5, 18% N, and 2% S) became popular, especially in developed countries, in the 1960s. Other sources of phosphate included direct application of phosphate rock, PAPR, nitrophosphates, and other compound fertilizers. Other sources of S included gypsum, which was rather popular as −land plaster− in the eighteenth century. However, the use of gypsum slowly disappeared with the development and use of chemical fertilizers.

The world trends in production capacity for N, P_2O_5, S-containing fertilizers, and S-free fertilizers are shown in Figure 7.3. While the world capacity to produce N and P_2O_5 fertilizers rapidly increased from 1967 onward, the percentage share of AS in total N, and of SSP in total P_2O_5 declined. For the world as a whole, the contribution of AS in total N production capacity declined from 18% in 1967 to 9% in 1982, whereas the share of urea increased from 24% in 1967 to 46% in 1982. As far as P_2O_5 world production capacity is concerned, the share of SSP declined from 43% in 1967 to 20% in 1982, whereas the share of TSP increased from 23% in 1967 to 27% in 1982. The share of AS and SSP declined not only in world production capacity, production, and consumption but also in world trade. For example, as was discussed in British Sulphur Corporation (1967, 1972), the relative share of AS in the world N exports declined from 33% in 1963/64 to 17% in 1969/70 while the corresponding share of urea increased from 22% to 41%.

The total world consumption of AS has been rather static. The estimated consumption of N from AS increased from 2.7 million mt in 1960/61 to 2.8 million mt in 1980/81, while the total N consumption increased almost six-fold from 10.2 million mt in 1960/61 to 60.3 million mt in 1980/81, resulting in a decline in the share of AS from 26% in 1960/61 to 5% in 1980/81. The relative contribution of AS in total N consumption in the world regions is

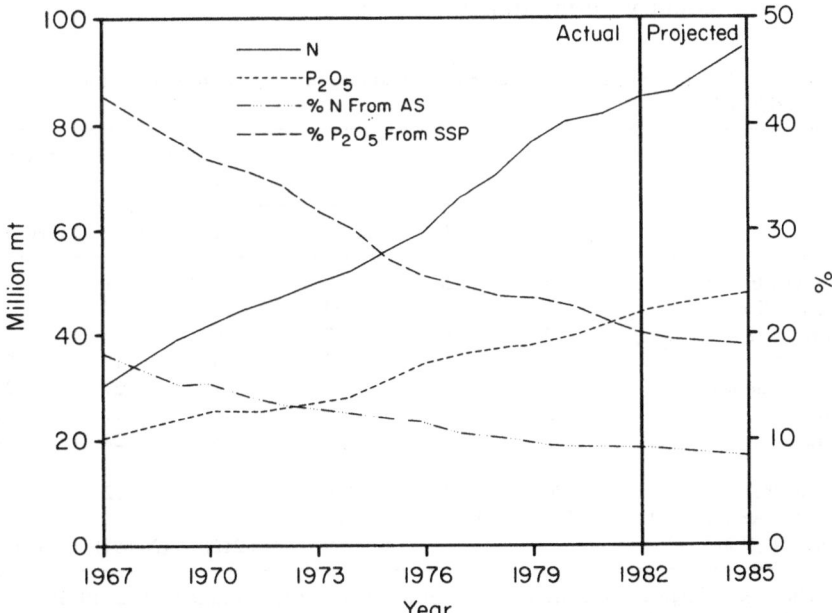

Figure 7.3. World Trends in Production Capacity for Sulfur-Containing Nitrogen and Phosphate Fertilizers.

140

reported in Table 7.8. Asia has always accounted for the largest share of AS consumption in the world. However, even for Asia, the relative share droppedfrom 42% in 1960/61 to 27% in 1980/81. The Latin American countries, particularly Brazil, are emerging as important consumers of AS.

Similarly, the relative share of SSP in total P_2O_5 fertilizer production capacity is reported in Figure 7.4. In 1982, the relative share of SSP was 7% in Africa (excluding Egypt, Libya, Sudan, and South Africa); 21% in Far East(excluding China and Japan); and 28% in Latin America. Finally, the relative contribution of AS to total N and of SSP to total P_2O_5 consumption in selected developing countries is reported in Table 7.9. With the exception of a few countries, the world supply of N and P_2O_5 is dominated by S-free fertilizers.

The amount of S supplied as plant nutrient by S-containing fertilizers is estimated as follows:

9. $S_{ijt} = \alpha_i F_{ijt}$, and

10. $S_{jt} = \Sigma_i S_{ijt}$.

Where

S_{jt} → Total amount (mt) of S supplied by chemical fertilizer in jth country during time t,

Table 7.8. Relative contribution of ammonium sulfate to regional nitrogen consumption in the world[a]. (%)

Region	Regional share in AS consumption			Contribution of AS to regional N consumption		
	1960/61	1969/70	1980/81	1960/61	1969/70	1980/81
Western Europe	26	20	15	22	11	4
Eastern Europe[b]	13	12	23	22	7	5
North America	10	12	7	9	6	2
Oceania	1	1	2	65	15	22
Latin America	6	9	21	37	27	21
Africa[c]	2	2	5	19	11	7
Asia[d]	42	43	27	68	22	4
World (%)	100	100	100	26	12	5
N from AS ('000 mt)	(2 701)	(3 454)	(2 828)			
Total N (' 000 mt)				(10 212)	(28 691)	(60 336)

a. The original figures were derived from British Sulphur Corporation (1972, 1983a).
b. Including U.S.S.R.
c. Continental Africa, including Egypt, Libya, Sudan, and South Africa.
d. Including Israel, Japan, and China.

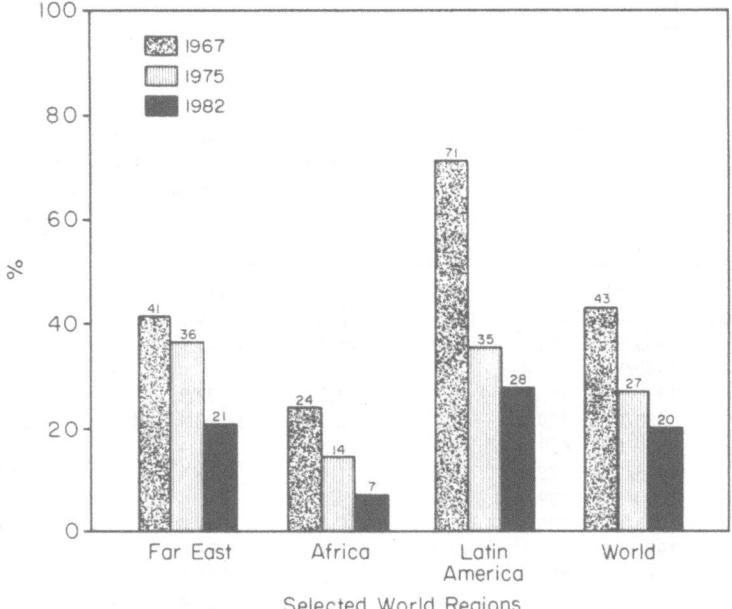

Figure 7.4. Relative Contribution of Single Superphosphate to Regional P_2O_5 Fertilizer Production Capacity in the World.

α_i → Sulfur supply coefficient (percent S) in ith fertilizer, and

F_{ijt} → Consumption (mt) of ith S-containing chemical fertilizer in jth country during time t.

The chemical fertilizers used in estimating S the supply include AS, ammonium sulfate nitrate, SSP, and potassium sulfate. There may be other fertilizers, especially compounds, and soil amendments such as gypsum that supply S. However, appropriate data were not available to estimate the S supply from these sources. Consequently, the total S supply is underestimated in terms of the S supplied by these sources. In any case, the amount of S supplied by these sources in the tropics may be rather small.

Appropriate data were not available from all the countries and regions for determination of the S supply; they were available for India and Brazil. The estimated S consumption for India and Brazil is reported in Table 7.10. The results indicate that in the 1950s and 1960s both India and Brazil did consume large amounts of S in comparison with N, P_2O_5, and K_2O. However, from the latter part of the 1960s onward the consumption of N, P_2O_5, and K_2O skyrocketed, and the consumption of S did not keep up. The primary reason for this pattern was the decline in the percentage contribution by AS and SSP to total N and P_2O_5 consumption, respectively.

Table 7.9. The relative contribution of ammonium sulfate and single superphosphate to nutrient consumption in selected developing countries[a].

Country	Year	Nitrogen (N)			Phosphate (P_2O_5)		
		Total N ('000 mt)	N from AS ('000 mt)	N from AS as % of total N (%)	Total P_2O_5 ('000 mt)	P_2O_5 from SSP ('000 mt)	P_2O_5 from SSP as % of total P_2O_5 (%)
Algeria	1980/81	89.7	0.4	0.4	114.9	–	–
Bangladesh	1980/81	266.7	b	b	119.9	–	–
Brazil	1980/81	886.0	210.0	23.7	1 965.6	350.5	17.8
Chile	1980/81	52.4	–	–	71.0	1.1	1.5
Colombia	1979/80	151.0	4.8	3.2	73.2	–	–
India	1980/81	3 529.7	103.7	2.9	1 090.4	170.9	15.7
Indonesia	1979/80	629.2	41.1	6.5	151.4	–	–
Kenya	1978/79	25.4	7.9	31.1	16.1	1.3	8.1
Malaysia	1980/81	139.4	6.3	4.5	118.8	–	–
Mexico	1981/82	1 106.3	332.1	30.0	369.8	73.3	19.8
Morocco	1981/82	81.1	17.9	22.1	78.9	17.4	22.1
Pakistan	1980/81	875.3	20.3	2.3	243.6	20.2	8.3
Peru	1981/82	101.7	4.8	4.7	21.4	1.6	7.5
Philippines	1980/81	233.7	25.1	10.7	52.8	0.5	0.9
Sri Lanka	1979/80	77.2	22.5	29.1	23.5	–	–

a. Original data were obtained from FAO (1983) for Colombia and Kenya and from ISMA (1982) for the other countries. The consumption data are first-order approximations since for some countries no distinction was made between consumption and distribution.
b. A small amount of AS is produced and used on tea plantations.
– None or negligible.

Estimated Gaps in Sulfur Requirements and Sulfur Supply

National gaps in S requirements and S supply are estimated as follows:

11. $G_{jt}^U = S_{jt} - U_{jt}.$

12. $G_{jt}^R = S_{jt} - R_{jt}.$

Three types of S gaps are estimated at the national level. These are:

Gap I $[G_{jt}^U]$ → Difference between S supply and S uptake.

Gap II $[G_{jt}^R(I)]$ → Difference between S supply and S replacement requirements with S replacement coefficient of 1.75.

Table 7.10. Estimated consumption of sulfur and other primary nutrients in India and Brazil.

Country	Year	Consumption of				N from AS as % of total N	P₂O₅ from SSP as % of total P₂O₅
		N ('000 mt)	P₂O₅ ('000 mt)	K₂O ('000 mt)	Sᵃ ('000 mt)		
India[b]	1956/57	108	13	10	130	96	99
	1960/61	212	53	29	224	69	99
	1965/66	547	132	78	395	48	81
	1970/71	1 310	305	198	346	16	36
	1975/76	1 909	374	227	227	8	20
	1980/81	3 522	1 074	618	250	3	16
Brazil[c]	1956/57	30	56	42	35	37	57
	1960/61	67	78	106	73	51	60
	1965/66	71	87	100	95	64	72
	1970/71	279	377	306	252	52	33
	1975/76	406	1 014	558	266	26	21
	1980/81	886	1 966	1 267	482	24	18

a. Sulfur contained in AS, ammonium sulfate nitrate, SSP, and potassium sulfate.
b. Original data were obtained from FAI (1983).
c. Original data were obtained from FAO (1983) up to 1978/79; and from ISMA (1982) for 1979/80 and 1980/81.

Gap III $[G_{jt}^R(II)]$ → Difference between S supply and S replacement requirements with S replacement coefficient of 3.50.

The estimated S gaps for selected tropical countries during 1980/81 are reported in Table 7.11. With the exception of Gap I in Brazil and Mexico and Gap II in Mexico, all three gaps are estimated to be negative, which implies that the S requirements are larger than the S supply. With the exception of Mexico, all the countries listed have no known resources of indigenous S.

A more detailed analysis of the S gaps is performed for India. The estimated S gaps are reported in Table 7.12. All the three S gaps (Gap I, Gap II, and Gap III), which serve as guides for developing S supply strategy, are negative, and the amount of S needed to bridge these gaps is rather large. For example, the estimated S Gap III for India during 1980 was 2.5 million mt, and it is expected to increase to 3.1 million mt in 1990 and 4.0 million mt in the year 2000.

The models developed and used in this study provide a systematic means of assessing S requirements for agricultural crops, S requirements for fertilizer industry, fertilizer S supply, and the potential S gaps that may prevail. While the results indicate that large S gaps now exist or soon will emerge in a number of developing countries, we wish to point out that these esti-

144

Table 7.11. Estimated sulfur requirements, fertilizer sulfur supply, and sulfur gaps in selected countries during 1980. ('000 mt of S)

Country	Sulfur requirements			Fertilizer sulfur supply[a]	Sulfur gaps[b]		
	Uptake	Replacement I	Replacement II		I	II	III
Asia							
India	784	1 372	2 743	250	− 534	− 1 122	− 2 493
Indonesia	130	227	454	48[c]	− 82	− 179	− 406
Philippines	45	80	159	30	− 15	− 50	− 129
Africa							
Kenya	15	26	52	11[d]	− 4	− 15	− 41
Niger	13	22	44	< 1[c]	− 12	− 21	− 43
Nigeria	66	116	231	17[c]	− 49	− 99	− 214
Sudan	30	53	107	[e]	[e]	[e]	[e]
Zimbabwe	16	29	58	[e]	[e]	[e]	[e]
Latin America							
Brazil	349	611	1 222	482	+ 133	− 129	− 740
Colombia	38	67	134	5[c]	− 33	− 62	− 129
Mexico	137	239	478	338	+ 201	+ 99	− 140

a. Derived from FAO (1983) and ISMA (1982).
b. Fertilizer supply minus sulfur requirements. Gap I is supply minus uptake; Gap II is supply minus Replacement I; and Gap III is supply minus Replacement II.
c. For 1979/80.
d. For 1978/79.
e. Not available.

Table 7.12. Estimated sulfur requirements, fertilizer sulfur supply, and sulfur gaps in India. ('000 mt of S)

Year	Sulfur requirements			Fertilizer sulfur supply[a]	Sulfur gaps[b]		
	Uptake	Replacement I	Replacement II		I	II	III
1960	509	891	1 782	224	− 285	− 667	− 1 558
1965	524	917	1 833	395	− 129	− 522	− 1 438
1970	651	1 139	2 279	346	− 305	− 793	− 1 933
1975	688	1 204	2 407	226	− 462	− 978	− 2 181
1980	784	1 372	2 743	250	− 534	− 1 122	− 2 493
1985	886	1 550	3 101	343	− 543	− 1 207	− 2 758
1990	1 008	1 764	3 528	417	− 591	− 1 347	− 3 111
1995	1 155	2 021	4 043	508	− 647	− 1 513	− 3 535
2000	1 333	2 332	4 665	618	− 715	− 1 714	− 4 047

a. Fertilizer sulfur supply for 1985 and onward is estimated by assuming 4% annual compound growth from a base of 271 000 mt S consumption for 1979, which is a 3-year (1978/79, 1979/80, 1980/81) simple average.
b. Fertilizer sulfur supply minus sulfur requirements. Gap I is supply minus uptake; Gap II is supply minus Replacement I; and Gap III is supply minus Replacement II.

mates are firstorder approximations. Although the S gap estimates are based on an analysis of the best available data, much work remains to be done for further refinements in the context of specific countries. Recognizing these limitations, the results have major implications for national fertilizer policy.

The large S requirements and gaps have important policy implications with respect to S research, S supply, fertilizer material selection, fertilizer imports, fertilizer distribution, S promotion, investment, and foreign exchange allocation. However, unless something is done to bridge these large S gaps and to correct the S-deficiency problem,national and international efforts to accelerate the domestic food and agricultural production in most of the tropical countries will be seriously handicapped.

8   Fertilizer Sulfur Sources and Supply Strategies

In order to correct the increasing S deficiencies and bridge projected S gaps, there is a need to identify, develop, evaluate, and transfer fertilizer S technology and strategies that would be appropriate, technically and economically, for tropical countries of the world. Alternative S supply strategies include conventional S-containing fertilizers, modified S-containing fertilizers, and indigenous S supply sources such as native S, gypsum, phosphogypsum, and pyrites. The purpose of this chapter is twofold: (1) to briefly discuss various S supply sources, including S-containing fertilizers and soil amendments and (2) to discuss appropriate S supply strategies in the context of developing tropical countries.

Sulfur Sources

As has been analyzed in the preceding chapters, the S status of the soil is improved through addition of S from precipitation, atmospheric dust, irrigation water, organic material, crop residues, fertilizers, and soil amendments.

Atmospheric Accession

Unlike the situation in highly industrialized areas, the atmospheric accession of S in the agricultural areas of most of the developing countries is rather low. According to Jones (1978), most of the atmospheric accession can be brought down by a precipitation of the first 15 mm of rainfall. Any more rain only dilutes the concentration of sulfate in leaching water, as has been shown by Bromfield, Debenham, and Hancock (1980) in Kenya. Most of the atmospheric S is expected to be added to soil after a prolonged dry season and with the first rain of the wet season. The net addition of S to soil, however, will depend on many factors, some of which have not been analyzed. The supply of S through dust and gaseous deposition on plants and soils can also be an important source of S, particularly in semiarid tropics and arid zones, since these areas have more dry periods and dust storms.

Irrigation Water

Irrigation water is an important source of S supply to soil. However, regular monitoring of the quality of water is essential to determine the potential supply of S from irrigation water. In irrigated areas of semiarid tropics and arid zones, the S supply from irrigation water can be adequate for heavy soils but probably not for coarse-textured and highly permeable sandy soils. In this context information on the residual buildup of sulfate, S leaching, and immobilization of S with the organic matter of the soil needs to be developed. Simulation models in the laboratories are no doubt useful, but more valuable results on S supply from irrigation water can be obtained only from field studies.

Organic Material and Crop Residues

The organic residues of the crop are important sources of S, and generally more S is retained in these residues than is removed by the grain or the marketable product of the crop. Incorporation of organic residues, farmyard manures, and other oganic residues in the soil can build up soil S reserves. However, because of rapid mineralization and excessive leaching inmany tropical soils, the buildup of sulfate through this mechanism is ratherlow, expecially in surface soil. Some of the S may move down in the soil profile and get adsorbed, thus adding to S reserves of the soil. The crop residue management is a practical problem that should receive adequate attention in any strategy to supply S and build up soil fertility. Approximately 1 mt of farmyard manure adds about 2 kg of S, and its availability is very low. Thus, unless a substantial amount of manure is added, the S supply through farmyard manure will be rather small. Moreover, it may also cause more imbalance between N and S supply. In most developing tropical countries, crop residues and straw are generally removed for use as fodder or fuel.

Fertilizers and Soil Amendments

The most important source of S to soil in the developing countries is through S-containing fertilizers and soil amendments. Substitution of S-free high-analysis fertilizers such as urea and TSP for AS and SSP is increasing the gap between S requirements and S availability. For example, 100-kg applications each of N, P_2O_5, and K_2O supplied from urea, TSP, and MOP, respectively, result in a decreased supply of approximately 217 kg of S when a switch is made from the corresponding S-containing fertilizers such as AS, SSP, and potassium sulfate.

Though the supply of S in the agricultural system of developing countries is declining, there are few deliberate efforts to reverse the trend through use of S-containing conventional or modified fertilizer products or by incorporation of S through other supply sources. In fact, most of the S that would have gone directly to soil as a part of the fertilizer is being thrown away as a waste product of the phosphate industry, and little attempt is being made to use it as a source of S in agriculture. Somehow the declining trend in S supply needs to be reversed. It can hardly be overstated that from the point of view of modernizing agriculture, it is essential to understand the problems and prospects of using S-containing substances to provide S to impoverished soils of tropical regions, cheaply and effectively.

Sulfur-Supplying Fertilizers and Soil Amendments

The S-containing fertilizers and amendments can be divided into six groups: (1) dry solid fertilizer products containing sulfate, (2) dry solid fertilizer products containing elemental S, (3) fluid products, (4) organic products, (5) compound fertilizers and mixtures, and (6) byproducts of industry. Each of these groups can be further subdivided into sulfate-containing, elemental S-containing, and complex substances. Table 8.1 and Appendix II give the S content of the important fertilizer-supplying substances. However, this is not an exhaustive list as there are numerous other grades and products that contain S.

The S content of S-containing fertilizers and substances varies considerably in amount and forms of S, i.e., sulfide-S, sulfate-S, and organic S. Forms of S other than sulfate-S must undergo mineralization or oxidation to sulfate before they become available to plants. Sulfur-oxidizing organisms of Thiobacillus species bring about such a conversion. The rate of conversion depends on a number of factors that are very well documented in textbooks of microbiology. However, the chemical changes in S under field conditions of tropical soils have not received adequate attention.

It is beyond the scope of this study to discuss the technology and properties of the S-containing substances. Bixby and Beaton (1970), Beaton and Fox (1971), and Hignett (1979) have dealt with this subject in some detail. However, the technical aspects of S-containing fertilizers need more attention from fertilizer technologists, agronomists, and economists.

Considerations for Formulating a Sulfur Supply Strategy

Each developing country must consider its own specific S problems and S supply sources in formulating a national S supply strategy. There are some broad considerations that must be kept in mind.

1. A large number of fertilizer products are available in the industrialized countries, and these countries are also able to manufacture products to meet specific needs. However, the developing countries generally use only a limited number of fertilizer products because of lack of availability and/or cost considerations. The proportion and total amount of each of these fertilizer products varies from country to country. A national policy to supply S as a nutrient should be developed on the basis of indigenous raw material, fertilizer response, fertilizer price, and crop price information.

150

Table 8.1. Sulfur-containing fertilizers and other substances[a].

Sulfur containing substances	Nutrient content (%)				N:S ratios	P:S ratios
	S	N	P	K		
I. Dry Fertilizer substances containing sulfate						
With S content more than 10%						
Ammonium sulfate	24	21	–	–	0.9	–
Ammonium sulfate nitrate[b]	15	26	–	–	1.7	–
Ammonium phosphate sulfate[b]	15	16	9	–	1.1	0.6
Potassium sulfate	16–22[c]	–	–	40	–	–
Magnesium sulfate monohydrate	23	–	–	–	–	–
Potassium magnesium sulfate[d]	22	–	–	18	–	–
Gypsum	18	–	–	–	–	–
Zinc sulfate	17–18	–	–	–	–	–
Single superphosphate	12	–	9	–	–	0.8
With S content less than 10%						
Superphosphate double	9	–	13	–	–	1.4
Diammonium phosphate	1[e]	18	20	–	14.0	15.0
Triple superphosphate	1.0	–	20	–	–	20.0
Phosphate rock	<1.0	–	14	–	–	>14.0
Urea ammonia sulfate	4–13[f]	40	–	–	3.1–10.0	–
Monoammonium phosphate	1	11	24	–	11.0	24.0
II. Dry fertilizers containing elemental S or sulfide						
Elemental S	100	–	–	–	–	–
Sulfur bentonite	90	–	–	–	–	–
Sulfur-fortified single superphosphate[b]	27	–	7	–	–	0.3
Phosphate rock-S	7–16	–	9	–	–	0.6–1.3
MOP-urea-S	g	–	g	–	–	g
Sulfur-coated urea	14	38	–	–	2.7	–
Pyrites	>40	–	–	–	–	–
Urea-S[b]	10	40	–	–	4.0	–
Sulfur-fortified concentrated superphosphate[b]	20	–	18	–	–	0.9
III. Fluid fertilizers containing S						
Ammonium thiosulfate solution	26	12	–	–	0.5	–
Ammonium polysulfide	40–50	20	–	–	0.4–0.5	–
Ammonium bisulfate solution	17	8.5	–	–	0.5	–
Ammonium bisulfate polyphosphate	3–5	9	8	–	1.8–3.0	1.6–2.7
Sulfur dioxide	50	–	–	–	–	–

Table 8.1. Continued.

Sulfur containing substances	Nutrient content (%)				N:S ratios	P:S ratios
	S	N	P	K		
IV. Organic S-containing substances, manures and composts						
Sewage sludge	0.4	–	–	–	–	–
Bonemeal	0.2	–	–	–	–	–
Groundnut meal	0.2	–	–	–	–	–
Farmyard manure (wet – 85% moisture)	0.2	0.3	0.2	0.3	–	–

V. Compound fertilizers and mixtures containing N, P, K, S, and micronutrients

There are numerous possibilities and, in fact, already many products are available.

VI. Dry solid byproducts of industries or natural ores

Anhydrite – 23.5% when pure
Gypsum – S content variable depending upon purity – 13% – 18% sulfate S
Phosphogypsum – byproduct of phosphoric acid or phosphatic fertilizer industry
Pyrites – S content ranging from 30% to 60% and above
Brimstone or other S ores – S content up to 100% (ores contain 25% – 100% S)
Pressmud of sulfitation process – byproduct of the sugar industry with S content highly variable
Pressmud of paper industry – byproduct of paper industry variable S content
Miscellaneous byproducts of agricultural industries

a. The conversion factors for plant nutrients from oxide to elemental and from elemental to oxide forms are: $P = 0.4364$ P_2O_5, $P_2O_5 = 2.2914$ P, $K = 0.8302$ K_2O, and $K_2O = 1.2046$ K.
b. Other grades are also marketed.
c. Average 18%.
d. Langbeinite.
e. 1.3%, approximately.
f. Average 6%.
g. Variable.

2. Although there are many S-containing fluid substances, it seems doubtful that the use of fluid fertilizers containing S will be practical in the near future in developing tropical countries.
3. The fertilizers are normally priced only for their N, P, or K content, and generally no consideration is given to their S content. Table 8.2 gives the total nutrient content (N + P_2O_5 + K_2O + S) of a few important fertilizers that contain high amounts of S, in addition to other major plant nutrients. The sulfate-containing dry fertilizer substances with high sulfate content may become more attractive provided the fertilizer pricing policy or transportation cost does not outprice them.

For example, AS has 21% N and 24% S. Thus, total nutrient content

Table 8.2. Estimated total nutrient content of selected nitrogen and potassium fertilizers: An example.

Fertilizer	S	Total plant nutrients[a] (%)
Ammonium sulfate[b]	24	45
Ammonium nitrate sulfate[c]	5	35
Ammonium phosphate sulfate[d]	15	52
Ammonium sulfate nitrate[e]	12	38
Potassium sulfate[f]	18	68
Potassium magnesium sulfate[g]	22	44

a. $N + P_2O_5 + K_2O + S$.
b. 21% N.
c. 30% N.
d. 16.5% N and 20.5% P_2O_5.
e. 26% N.
f. 50% K_2O.
g. 22% K_2O.

in AS is 45%, whereas urea has 46% N. If AS is priced on the basis of N content, taking urea N price as standard, AS is far more economical than urea if it is used on S-deficient soils. Sulfur in this case is a bonus. However, if pricing policy gives positive value to S, ammonium sulfate may be outpriced in comparison with urea as a source of N. There may be situations where AS could be considered as a major source of S and N as a bonus. Such decisions can be made only on a sound technological, agronomic, and economic basis. On the other hand, the fertilizer industry will have limited economic incentive to supply S if consideration is not given to S in pricing fertilizers. It is important to point out, however, that the cost of S if used in producing fertilizers is incorporated in fertilizer manufacturing cost by the fertilizer industry.

The importance of S in pricing fertilizers will depend on its response ratio in comparison with N, P, or K. It is doubtful whether any developing country has a pricing policy based on the total nutrient content. In some states in the United States the fertilizer legislation requires that all mixed fertilizers should contain at least 3% S. Similar policies may be needed for developing countries faced with S-deficiency problems.

Some fertilizers containing not less than 5% (5%-10%) sulfate-S, ncluding PAPR, can become attractive for certain crops in areas having S deficiency. Grant and Rowell (1976) suggested a 6.5% S content for fertilizers and a ratio of P_2O_5:S of 2.1-2.7:1 for Zimbabwe. Bixby and Beaton (1970) have suggested a ratio of 3:1. However, the appropriateness of these ratios needs to be investigated in the context of tropical soils and cropping systems.

4. Of the fertilizer substances containing elemental S, S-coated urea, S-fortified superphosphate, and phosphate rock S can become attractive fertilizers and, hence, sources of S for tropical soils.

5. Elemental S and pyrites will be preferable for alkaline and calcareous soils, provided they are used in such a way that rapid oxidation is facilitated and the cost-benefit considerations are favorable. Gypsum is useful on both alkaline and acid soils.

6. In soils of low pH, the use of elemental S or related substances and AS may not be so advisable because of their acidifying effect, but in calcareous and alkaline soils they have an advantage provided the cost:benefit ratios justify their use.

7. Undoubtedly, the use of organic sources of S such as crop residues and other sources is desirable, but it will take a long time to correct the S-deficiency problem of soil if it can be corrected at all. Moreover, the amount of manure available for extensive use in tropical countries is rather limited. Crop residues are generally used as fodder or fuel.

8. The use of byproducts of the chemical fertilizer and agricultural-based industry as S supply sources is very promising and needs intensive research. However, since this research requires a considerable specificity of location, every country has to formulate a research strategy according to the availability of these products and their economic value.

    In India, for example, the use of low-grade gypsum, phosphogypsum, low-grade pyrites, pressmud of sugar and refuse from paper factories offers great promise for correcting the S deficiency in cultivated soils and for reclamation of sodic soils. As the phosphate fertilizer industry develops, a large amount of gypsum as a byproduct of this industry will be available. Agarwal (1982) has estimated that by the end of the sixth 5-year plan, India will annually accumulate more than 5 million mt of phosphogypsum as a byproduct of the phosphate industry. Bangladesh is also faced with a similar problem of large stocks of phosphogypsum.

    There are two possibilities for using phosphogypsum, one in agriculture and another in industry. So far as agriculture is concerned, phosphogypsum can be used as a nutrient source as well as a soil amendment. A few developing countries, including Bangladesh, have already started using phosphogypsum in agriculture but only in small amounts. Some of the S from this source can be reintroduced into fertilizer products for enriching them with S. However, this problem needs to be addressed jointly by technologists, agronomists, and economists. Ultimately, the use of phosphogypsum, whether in agriculture or industry, depends upon economic considerations.

9. Sulfur-coated urea, PAPR, S-fortified TSP, double superphosphate, and S- or sulfate-enriched DAP need to be critically evaluated for use in S-deficient soils and crops. In fact, S-coated urea may offer an advantage for supplying S to S-deficient soils, in addition to improving N

efficiency. Sulfur may not have been given due consideration in research on testing S-coated urea in developing countries. Likewise, studies on the value of PAPR not only as a source of more soluble phosphates but also as a source of S, particularly for S-deficient soils or for S-responsive crops, need to be expanded. Ammonium sulfate needs to be given a fresh look by the technologists, agronomists, and policymakers, especially in the tropics.

10. The use of zinc sulfate, which is increasing in India and other Asian countries for correcting Zn deficiency, also should be exained from the point of view of correcting S deficiency. Every 50 kg of zinc sulfate, while supplying approximately 18 kg of Zn, also supplies about 8 kg of sulfate-S.

Modification of Fertilizers to Supply Sulfur

Fertilizer technologists need to give considerable attention to the modification of popular fertilizers to supply S. In fact, there appear to be numerous possibilities for such modifications. However, the choice will depend on a number of factors such as ease of application, agronomic effectiveness of products, economics, need for multinutrient products, and the availability of advisory and marketing services in the country.

The S fertilization strategy will vary from country to country. However, any S supply strategy should consider (1) nature of soils, (2) nature of crops, (3) cropping system, (4) major nutrient deficiencies and their relationship to S deficiency, (5) soil fertility management system, (6) use of organic manures, (7) expected yield level, (8) socioeconomic factors, (9) pricing of fertilizers and crops produced, (10) benefit:cost ratios, and (11) availability of indigenous raw materials.

For most situations, the following products offer great promise: gypsum, ammonium phosphate sulfate, ammonium sulfate nitrates, superphosphates, acidulated phosphate rock, and fertilizer mixtures and compound fertilizers supplying S in addition to major plant nutrients. Tisdale and Platou (1981) consider that the following high-analysis S-containing fertilizers hold some promise: (1) DAP + S, (2) TSP + S, (3) S bentonite, (4) urea-S, (5) S-coated urea, (6) urea-AS, and (7) ammonium nitrate sulfate. Joint efforts of fertilizer technologists, agronomists, and economists are needed to develop suitable products for different situations in tropical countries.

Research on Sulfur Fertilizers

In view of the growing concern about S deficiency in tropical agriculture

and its implications for fertilizer technology and management, IFDC initiated a research program on S which has emphasized the following aspects:
1. Study of the effectiveness of modified phosphates and nitrogenous fertilizers through incorporation of gypsum and elemental S.
2. Study of the causes and mechanism of losses of S in tropical soils and means of correcting them through use of S-containing substances.

Modified Phosphate-Sulfur Fertilizers

The incorporation of gypsum in high-analysis phosphates such as TSP + gypsum and DAP + gypsum and the use of PAPR have been investigated under greenhouse conditions using S-deficient soil and corn as a test crop. The results show that all the phosphates with incorporated gypsum ranked significantly better than TSP as judged by dry-matter production and S uptake and compared favorably with SSP (IFDC Annual Report 1981). The studies also indicate that cogranulation or blending of TSP with gypsum so as to give a ratio of P:S of 3:1 was best for supplying both nutrients (Korentajer, Mokwunye, and Hellums, 1982). Collaborative field trials in Kenya, Nigeria, and Burkina Faso using PAPR also show the effectiveness of this source for supply of P and S.

Modified Nitrogen-Sulfur Fertilizers

The nitrogenous S-containing fertilizers studied at IFDC were(1) gypsum-coated urea, (2) gypsum plus urea, (3) powdered urea plus elemental S, and (4) urea-elemental S granules. The test material was applied by banding at 8-cm depth or by broadcasting on the surface of the soil. The highest dry-matter production and S uptake were with gypsum-coated urea ad the lowest with urea-elemental S (IFDC Annual Report 1981).

These studies show that nitrogen and phosphate fertilizers may be modified by incorporating gypsum to become effective and economical sources of S to plants that are suffering from an S deficiency. In order to study the practical feasibility of these promising products, however, field experiments in tropical African, Asian, or Latin American countries are necessary.

Replenishing Sulfur Lost Through Leaching

From leaching experiments it was observed that apparent amounts of gypsum removed by leaching varied from 8% to 58% of the amount added; the lowest leaching rate was 0.91 mm/day, and the highest was 4.2 mm/day.

To replenish the S lost through leaching, sources such as gypsum, elemental S, and other S-containing substances are commonly used. A study was conducted to find the particle size, rate of application, and method of

156

application of these substances that would be appropriate for correcting S deficiency or replenishing S losses.

Three sources of S — gypsum, anhydrite, and elemental S — were compared at two rates of application — 10 and 40 ppm. The sulfate particles were of three sizes: (1) powder (-150 mesh), (2) 1.7-2.4 mm, and (3) 3.4-6.4 mm. They were compared with elemental S powder of one size only (<0.08 mm).

Before planting corn, the S-treated soil was subjected to a leaching rate of 1.9 mm/day for 10 days. The S losses were highest with the gypsum powder and decreased as the particle size increased. The amount of S leached varied from 1.8% to 67% of the amounts applied. In general, the losses were gypsum > anhydrite > elemental S (Figure 8.1).

These results suggest the use of coarser material instead of powder. Secondly, less-soluble material is preferable for reducing S losses. Sulfur-containing substances like anhydrite, pyrites, and elemental S may be preferable to gypsum in soils subject to high leaching losses; the question is one of economics, however, so it is desirable to conduct such experiments under field conditions to examine their direct and residual effects and the related economic aspects.

Criteria for Evaluating Fertilizer Sulfur Sources

In developing and recommending any of the S-containing fertilizers, it is extremely important to keep in mind the climatic conditions, cropping sys-

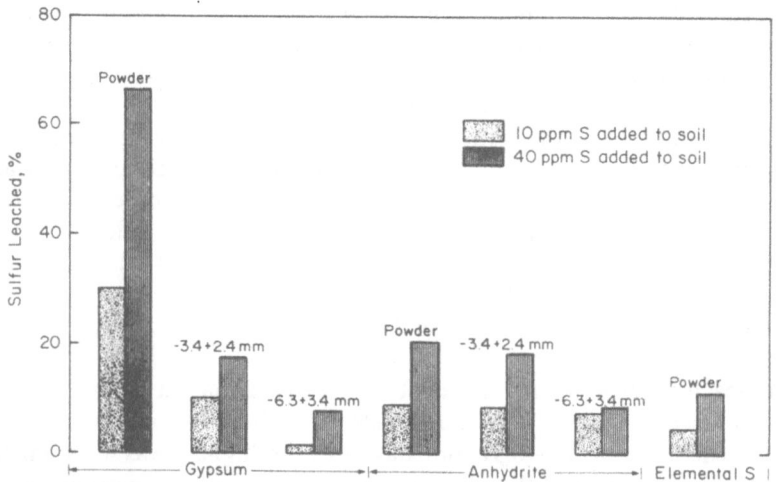

Figure 8.1. The Effect of Source, Particle Size, and Sulfur Rate on Sulfate Leaching Losses From Fertilizer Materials.

tems, soil types, socioeconomic conditions, and natural resource endowments of those tropical countries that are the target of such technology. All the existing and modified S-containing fertilizer technologies must be evaluated with respect to the following criteria:

1. The technical feasibility of production, distribution, and use of S-containing fertilizers.
2. Agronomic effectiveness under farmers' field conditions.
3. Preferences and general attitudes of those involved in production, distribution, and use of these materials.
4. Economic effectiveness under free-market conditions.
5. Economic effectiveness under prevailing and alternative government policies with respect to fertilizer S.
6. Foreign exchange use, earnings, and savings.
7. Economic and financial aspects of research, production, distribution, and the use of these materials.
8. Existing and suggested government policies dealing with fertilizer S raw materials, production, distribution, promotion, regulation, pricing, subsidies, trade, and research.

The technology for each proposed S-containing fertilizer needs to be evaluated within an interdisciplinary context from the moment a technology is conceived until it is ready for transfer and general use by farmers in the tropical countries.

9 World Sulfur Situation, Outlook, and Public Policy

Results reported in the preceding chapters clearly establish the strategic importance of S in economic development. Sulfur and its derivatives are needed to modernize the agricultural sector and also for industrial growth. The purpose of this chapter is threefold: (1) to examine the current economic situation and past performance of the S industry; (2) to analyze the economic outlook for S resources; and (3) to discuss public policies needed to ensure that adequate S is available for use by future generations. The analysis will deal with supply, demand, trade, and prices of S and its derivatives in the context of selected countries and regions of the world.

Sulfur Supply and Demand Components

Depending on the supply source, S is broadly classified into three categories: brimstone, pyrites, and S from other sources. Alternatively, S can also be considered as (1) natural and (2) involuntary or recovered.

Brimstone or elemental S consists of (a) Frasch − S obtained through Frasch mining, (b) native − S obtained by conventional mining, beneficiation, and refining, and (c) recovered − S recovered from sour natural gas and oil where desulfurization units provide hydrogen sulfide to plants that recover S in elemental form.

Pyrites, on the other hand, consist of (a) mining of pyrite (ferrous sulfide) ores and (b) byproduct pyrites derived from the smelting of nonferrous sulfide ores including copper, lead, nickel, and zinc.

Finally, S-in-other-forms consists of S obtained or recovered in the form of S ore, sulfuric acid, or other S derivatives. Some of these sources are SO_2 recovery from copper, lead, nickel, and zinc smelters; hydrogen sulfide from oil refineries; gypsum or anhydrite; and unbeneficiated S ore used in agricultural or industrial sectors.

A major portion of S is used to manufacture sulfuric acid which is considered the 'work horse' of industry. The S value of sulfuric acid is generally not retained in the final product. For example, in wet-process phosphoric acid manufacture, S from sulfuric acid is discarded in the byproduct phosphogypsum. The relative importance of various S uses in the United States is reported in Table 9.1. Agriculture, mainly the fertilizer industry, is by far the major consumer (62%) of S. The relative importance of agriculture as an S consumer is expected to decline to 40% by the year 2000. Potential new uses of S are expected to be S-asphalt paving and S concrete. At this stage, however, the economics does not appear to favor the use of S in concrete and paving material. Furthermore, S-imporing countries may never use S in building material and construction, unless phosphogypsum is used for

Table 9.1. Patterns of sulfur demand in the United States: An example[a].

End use of sulfur	1977 (actual)		2000 (forecast)[b]	
	'000 mt	% of total	'000 mt	% of total
Agriculture (fertilizers)	7 216	62.0	12 500	40.3
Plastic and synthetic products	455	3.9	1 200	3.9
Paper products	345	3.0	400	1.3
Paints	324	2.8	–	–
Metal mining and processing	701	6.0	1 800	5.8
Petroleum refining	879	7.5	2 000	6.5
Iron and steel production	130	1.1	–	–
Others[c]	1 607	13.8	3 600	11.6
Potential new uses[d]	–	–	10 000	32.3
Total[e]	11 657	100	31 000	100

a. Derived from Shelton (1979).
b. Forecasts are based on historical data and other economic indicators. The figures reported are the most probable. The forecast range for total S demand for the United States is 24.3 – 53.0 million mt.
c. Basically for chemical products.
d. Mainly S-asphalt paving and S concrete.
e. Totals are approximate due to rounding of data.

this purpose in countries with facilities for manufacturing wet-process phosphoric acid.

Patterns and Trends in Sulfur Production

World production of S increased from 37.8 million mt in 1969 to 53.8 million mt in 1981. The implied average annual growth in S production was 3.2% from 1970 to 1979 (3-year averages centered on years specified). There was a slight decline in S production during 1981 over 1980, mainly because of a fall in production in Iran and Iraq, large existing inventories, and slackening demand for S. However, Fertilizer Economic Studies (FERTECON) projects that world production of S will grow annually at 4% through the mid-1980s and slow to 3% in the latter part of the 1980s. As a result, the world S supply is expected to reach 62.8 million mt in 1985 and 73.0 million mt in 1990. The distribution of S production in individual countries and world regions and S sources during 1981 are reported in Table 9.2.

First, for the whole world, the share of brimstone (Frasch and recovered elemental S) in 1981 was 63%, a slight increase over 62% during 1973. The share is expected to increase to 68% during 1990. This increase is mainly attributed to an increase in recovered S in response to environmental protection regulations. The share of pyrite in 1981 was 21% and is expected to decline mainly in response to an increase in recovered (involuntary) S.

Second, the sources of S vary a great deal across countries and regions. For example, 90% of the S production in Latin America is from brimstone, 61% in Africa is from pyrites, and 40% in Asia is from other sources. These differences are due to several factors, including the existence of local S resources, level of industrialization, government regulations with regard to S emissions, and the relative economics of S production.

Third, the share of North America in total S production has declined from 40% in 1973 to 36% in 1981, whereas the share of the centrally planned economies has increased from 30% in 1973 to 35% in 1981. The relative changes in other world regions are rather small. No major changes are expected in the current production trends.

Fourth, during 1981 four countries accounted for 63.5% of total world· production of S: the United States (23.7%), U.S.S.R. (18.0%), Canada (12.6%), and Poland (9.2%). Changes in production in any one of these countries can have important implications for international S trade and S prices.

Fifth, with the exception of Mexico, the developing market economies produce a very small share of the world S output. During 1981 the estimated share of developing market economies in world S production was only 7.8%, and Mexico alone accounted for about 50% of that production. The oil-producing developing countries, especially in the Middle East, are expected to increase their production.

Patterns and Trends in Sulfur Consumption

World consumption of S has increased from 22.4 million mt in 1960 to 35.1 million mt in 1969 and 55.0 million mt in 1980. The average annual growth in S consumption from 1970 to 1979 (3-year averages centered on years shown) has been estimated to be 4.4%. The growth in derived demand for S has slowed down recently, mainly in response to the worldwide economic recession, high interest rates, and a slump in wet-proces phosphoric acid production.

Future demand for S depends on (1) S demand for fertilizer and agricultural uses, (2) S demand for current industrial (nonfertilizer) uses, and (3) S demand for potential new industrial uses. The available projections of worldwide demand for S-in-all-forms are summarized below:

Table 9.2. Production of sulfur by countries and world regions during 1981[a].

Region/country	Production ('000 mt)	Share in world production (%)	Sources of sulfur, % share		
			Brimstone	Pyrites	Other forms
Western Europe	7 691	14.31	47	32	21
Finland	455	0.85	10	52	38
France	2 117	3.94	93	0	7
Germany, F.R.	1 832	3.41	60	14	26
Italy	539	1.00	14	47	39
Norway	266	0.50	3	77	20
Spain	1 211	2.25	2	89	9
Sweden	290	0.54	14	62	24
Others	981	1.83	36	28	36
North America	19 520	36.33	85	3	12
Canada	6 775	12.61	87	5	8
United States	12 745	23.72	83	2	15
Oceania	153	0.28	9	0	91
Australia	153	0.28	9	0	91
Africa	814	1.52	7	61	32
Morocco	26	0.05	0	100	0
South Africa	636	1.18	8	71	21
Zambia	81	0.15	0	0	100
Others	71	0.13	14	32	54
Asia	4 041	7.52	50	10	40
India	195	0.36	3	21	76
Iran[b]	6	0.01	100	0	0
Iraq[b]	145	0.27	100	0	0
Japan	2 706	5.04	38	11	51
Kuwait	110	0.20	100	0	0
Philippines	70	0.13	0	100	0
Others	809	1.51	89	0	11
Latin America[c]	2 699	5.02	90	3	7
Chile	106	0.20	66	3	31
Mexico	2 167	4.03	95	0	5
Venezuela	45	0.08	100	0	0
Others	381	0.71	72	17	11
Eastern Europe	6 215	11.57	72	12	16
Poland	4 953	9.22	96	0	4
Romania	425	0.79	4	82	14
Yugoslavia	341	0.63	2	36	62
Others	496	0.92	26	51	23

Table 9.2. Continued.

Region/country	Production ('000 mt)	Share in world production (%)	Sources of sulfur, % share		
			Brimstone	Pyrites	Other forms
Other centrally planned	12 596	23.44	35	50	15
U.S.S.R.	9 670	18.00	40	42	18
China	2 591	4.82	18	79	3
Others	335	0.62	4	93	3
Western world	34 916	64.99	71	12	17
Centrally planned[d]	18 811	35.01	49	38	13
World[e]	53 727	100.00	63	21	16

a. Original production figures are obtained from British Sulphur Corporation (1982a, 1982b). However, all the calculations are by the authors.
b. Sulfur production has declined because of war between Iran and Iraq. During 1979 production of sulfur-in-all-forms was 244 thousand mt in Iran and 762 thousand mt in Iraq.
c. Excluding Cuba.
d. Sum of Eastern Europe and Other Centrally Planned Economies.
e. Totals are approximate due to rounding of data.

Source	Projected sulfur Demand (million mt)			Annual growth rate
	1985	1990	2000	
Shelton (1979)	76.0	NA	138.5	4.4%
Bixby (1980)	76.1	NA	NA	NA
BSC (1980)[a]	66.0	76.0	NA	3.5%, 1979 – 85 3.0%, 1985 – 90
'Markets Newsletter' (1982)	63.5	74.9	NA	<3.0%, early 1980s >3.0%, late 1980s

Clearly, there are large variations in S demand across different projections depending on the underlying assumption. Given the current state of the worldwide economy, the S demand in 1985 is expected to be in the neighborhood of 65 million mt.[1] According to Shelton and Morse (1983), total

1. According to Shelton (1979), cumulative worldwide demand for S is projected to be 510 million mt from 1977-1985 and 2,050 million mt rom 1977-2000.

world demand for S for industrial uses alone in 1990 is estimated to be 37.6 million mt, of which 84% is for current industrial uses and 16% is for potential new industrial uses.

Total S consumption in individual countries and regions is reported in Table 9.3. During 1980, 70% of the world S consumption was in the western world and 30% in the centrally planned economies. The share of individual countries in world S consumption during 1980 was 26% in the United States, 18% in the U.S.S.R., and 5% in Japan, accounting for approximately one-half of the world's consumption. The share of developing market economies in total S consumption was only about 12% during 1980, a rather small amount in the context of their relative share in world population and agricultural production.

The percentage share of brimstone in nonacid (for something other than sulfuric acid) uses varies from 3% in Africa to 15% in North America. During 1980, 12% of the brimstone in the western world was used for nonacid purposes. The rest of the brimstone was used for manufacturing sulfuric acid. On the other hand, most of the S from pyrites and from other sources was used to manufacture sulfuric acid either as the main product or as a byproduct. On the average, 88% of the total brimstone consumed in the western world during 1980 was allocated to the manufacture of sulfuric acid. Such information for the centrally planned countries was not available, but it is thought that the average percentage share allocated for production of sulfuric acid is not much different from that in the western world.

Patterns and Trends in Sulfuric Acid Production

Depending upon the country, 80%-90% of S is used to manufacture sulfuric acid. There are at least four reasons for this high figure: (1) sulfuric acid is cheapest among all the mineral acids; (2) sulfuric acid is quite versatile in its applications and ordinarily there is no satisfactory substitute; (3) the production of sulfuric acid results in net energy export for other uses; and (4) sulfuric acid is produced as a byproduct (e.g., smelting operations) or as a useful product from S derivatives recovered in response to government regulations to remove S emissions for environmental protection.

Sulfuric acid is produced by catalytic oxidation of SO_2 to sulfur trioxide. The sulfur trioxide is then absorbed in water or sulfuric acid to form sulfuric acid.[2] There are many different sources of SO_2 and alternative processes to convert SO_2 into sulfuric acid. Various sources of S for sulfuric acid production are (1) elemental (Frasch or recovered) S, (2) pyrite (ferrous sul-

---

2. Further details on the technical and energy-related aspects of sulfuric acid production are available in IFDC (1979), Mudahar and Hignett (1982).

Table 9.3. World consumption of sulfur and allocation of brimstone for nonacid uses during 1980[a].

Region/Country	Total consumption		Brimstone consumption for nonacid uses		% used for sulfuric acid
	'000 mt	% share	'000 mt	% of total S consumption	
Western Europe	11 016	20.0	1 254	11	89
Belgium	822	1.5	80	10	90
Finland	442	0.8	77	17	83
France	1 929	3.5	204	11	89
Germany, F.R.	1 701	3.1	182	11	89
Greece	450	0.8	48	11	89
Italy	1 263	2.3	167	13	87
Netherlands	532	1.0	NA	NA	NA
Spain	1 295	2.4	148	11	89
Sweden	298	0.5	60	20	80
United Kingdom	1 249	2.3	120	10	90
Others	935	1.7	NA	NA	NA
North America	15 919	28.9	2 319	15	85
Canada	1 899	3.4	370	19	81
United States	14 020	25.5	1 949	14	86
Oceania	1 054	1.9	70	7	93
Australia	805	1.5	55	7	93
New Zealand	249	0.5	15	6	94
Africa	3 192	5.8	106	3	97
Morocco	681	1.2	NA	NA	NA
South Africa	1 356	2.5	65	5	95
Tunisia	715	1.3	11	2	98
Others	440	0.8	NA	NA	NA
Asia	5 012	9.1	563	11	89
India	1 076	2.0	150	14	86
Japan	2 512	4.6	284	11	89
South Korea	562	1.0	40	7	93
Others	862	1.6	NA	NA	NA
Latin America[b]	2 488	4.5	253	10	90
Brazil	1 044	1.9	100	10	90
Mexico	951	1.7	65	7	93
Others	493	0.9	NA	NA	NA
Centrally planned[c]	16 367	29.7	NA	NA	NA
Eastern Europe	3 241	5.9	NA	NA	NA
U.S.S.R.	9 853	17.9	NA	NA	NA
Other centrally planned	3 273	5.9	NA	NA	NA
Western world	38 680	70.3	4 566	12	88
Centrally planned[c]	16,367	29.7	NA	NA	NA
World[d]	55 047	100.0	NA	NA	NA

a. Original consumption figures were obtained from British Sulphur Corporation (1982a, 1982b). However, all the calculations were made by the authors.
b. Excluding Cuba.
c. Sum of Eastern Europe, U.S.S.R., and Other Centrally Planned Economies.
d. Totals are approximate due to rounding of data. NA = Not available.

fide), (3) smelter operations (nonferrous sulfides), and (4) natural gypsum or byproduct phosphogypsum. On the average, 1 mt of sulfuric acid manufacture requires 330 kg S, all of which is contained in the final product (33% S in H_2SO_4). Furthermore, sulfuric acid manufacture is a source of useable energy in the amount of 1.32 GJ/mt of H_2SO_4 (1.25 million Btu/mt of H_2SO_4). The amount of useful energy generated, however, depends on the source of raw material S.

World production of sulfuric acid during 1981 was 137.8 million mt, a decline of 3.8 million mt over the previous year. On the average, however, production of sulfuric acid grew at 4.4%/year from 1974 to 1979 (3-year averages centered on years shown). A detailed analysis of sulfuric acid

Table 9.4. Production of sulfuric acid by country and world regions during 1981[a].

Region/country	Production ('000 mt)	Share in world production (%)	Sources of raw material, % share			All[c]
			Brimstone	Pyrites	Other forms	
Western Europe	25 541	18.5	56	27	17	100
Belgium	2 035	1.5	66	13	21	100
France	4 135	3.0	89	0	11	100
Germany, F.R.	4 220	3.1	48	18	34	100
Italy	2 500	1.8	44	39	17	100
Netherlands	1 798	1.3	85	0	15	100
Spain	2 993	2.2	0	90	10	100
United Kingdom	2 889	2.1	94	0	6	100
Others	4 954	3.6	38	43	19	100
North America	39 345	28.5	78	5	17	100
Canada	4 030	2.9	42	25	33	100
United States	35 315	25.6	82	3	15	100
Oceania	2 515	1.8	83	0	17	100
Australia	1 975	1.4	78	0	22	100
New Zealand	540	0.4	100	0	0	100
Africa	8 903	6.5	75	16	9	100
Morocco	2 353	1.7	97	3	0	100
South Africa	3 230	2.3	50	38	12	100
Tunisia	2 220	1.6	100	0	0	100
Others	1 101	0.8	55	12	33	100
Asia	12 718	9.2	56	7	37	100
India	2 780	2.0	90	4	6	100
Japan	6 572	4.8	25	11	64	100
South Korea	1 300	0.9	79	0	21	100
Taiwan	900	0.7	100	0	0	100
Others	1 166	0.8	93	7	0	100

production by region, country, and S supply sources for 1981 is given in Table 9.4. Approximately 69% of the world-021-s sulfuric acid was produced in the western world. The major producers include the United States (26%), U.S.S.R. (17%), China (6%), and Japan (5%). Among the developing market economies, major producers of sulfuric acid are India (2%), Mexico (2%), Brazil (1.7%), Morocco (1.7%), and Tunisia (1.6%). The developing marketing economies as a group account for 13.1% of the world's sulfuric acid production – Africa 4.2%, Asia 4.4%, and Latin America 4.5%.

The S supply sources for sulfuric acid vary across countries and regions. For the world as a whole, approximately 61% of the sulfuric acid is based on brimstone, 21% on pyrites, and 18% on other sources. Among the major

Table 9.4. Continued.

Region/country	Production ('000 mt)	Share in world production (%)	Sources of raw material, % share			All[c]
			Brimstone	Pyrites	Other forms	
Latin America	6 217	4.5	87	3	10	100
Argentina	275	0.2	80	0	20	100
Mexico	2 750	2.0	87	0	13	100
Brazil	2 400	1.7	93	7	0	100
Chile	385	0.3	71	3	26	100
Others	407	0.3	77	0	23	100
Eastern Europe	9 561	6.9	59	23	18	100
Czechoslovakia	1 350	1.0	100	0	0	100
Germany, D.R.	950	0.7	74	11	15	100
Poland	2 776	2.0	80	0	20	100
Romania	1 835	1.3	46	44	10	100
Yugoslavia	1 100	0.8	0	41	59	100
Others	1 550	1.1	33	54	13	100
Other centrally planned	33 060	24.0	39	46	15	100
U.S.S.R.	24 000	17.4	43	37	20	100
China	7 730	5.6	26	71	3	100
Others	1 330	1.0	35	62	3	100
Western world	95 240	69.1	70	12	18	100
Centrally planned[b]	42 621	30.9	43	41	16	100
World[c]	137 861	100.0	61	21	18	100

a. Original production figures are obtained from British Sulphur Corporation (1982a, 1982b). However, all the calculations were made by the authors.
b. Sum of Eastern Europe and Other Centrally Planned Economies.
c. Totals are approximate due to rounding of data.

sulfuric acid producers, the contribution of major S supply sources consists of 82% from brimstone in the United States, 43% from brimstone in the U.S.S.R., 71% from pyrites in China, and 64% from other sources in Japan. Consequently, the average production cost of sulfuric acid varies among these countries. Primarily, it is brimstone that moves in the international market. Other sources are either location-specific or too bulky for transportation.

Patterns and Trends in Sulfuric Acid Consumption

World consumption of sulfuric acid during 1980 was 142.8 million mt. On the average sulfuric acid consumption grew at 4.3%/year from 1974 to 1979 (3-year averages centered on years shown). Since demand for sulfuric acid is closely linked with demand for wet-process phosphoric acid, growth in future demand for sulfuric acid depends on the technical feasibility and economic viability of alternative processes for producing water-soluble phosphate fertilizers.

Regional consumption of sulfuric acid and its allocation to the fertilizer sector during 1980 are reported in Table 9.5. Almost one-half of all the sulfuric acid is consumed in Western Europe and North America. North America, especially the United States, is a major consumer of sulfuric acid, primarily because of its large and well-established phosphate fertilizer industry. The only developing countries that consume sulfuric acid in any significant amount are those that possess domestic capacity to produce phosphate fertilizers. To a large extent, domestic capacity to produce phosphate fertilizers is determined by the availability of phosphate rock.

The proportion of sulfuric acid used in the fertilizer industry varies from one world region to another. For the world as a whole, 56% of the sulfuric acid is used in the fertilizer industry. The percentage share varies from ashigh as 83% in Africa to a low of 44% in the U.S.S.R. and Western Europe. The rest of the sulfuric acid is used for other chemical industries. A large share of the sulfuric acid used in the fertilizer industry is used to manufacture phosphate fertilizers. For the western world, 91% of the fertilizer sulfuric acid is used in the phosphate fertilizer industry. The percentage share varies from a low of 73% in Asia to 99% in Oceania. The rest of the sulfuric acid is used to produce S-containing nitrogen fertilizer and other compound fertilizers.

The proportion of sulfuric acid used in the fertilizer industry has been changing over time. For the western world as a whole, this share increased from 51% in 1974 to 55% in 1979 and 58% in 1980. A major upward shift has occurred in North America where the share of sulfuric acid used in the fertilizer industry has increased from 46% in 1974 to 60% in 1979 and 62% in 1980. As far as other regions are concerned, the share has declined in Lat-

Table 9.5. World consumption and regional allocation of sulfuric acid and sulfur for fertilizer production during 1980[a].

Region	Total consumption		Consumption in the fertilizer sector			% share of sulfur used in fertilizer sector[b]	% used for phosphate fertilizers[c]
	Million mt	% share	Million mt	% share	% allocated to fertilizer sector		
Western Europe	27.2	19.0	12.0	15.0	44	39	86
North America	44.1	30.9	27.5	34.5	62	53	97
Oceania	2.9	2.0	2.3	2.9	79	73	99
Africa[d]	8.7	6.1	7.2	9.0	83	81	97
Asia[d]	12.2	8.5	6.4	8.0	52	46	73
Latin America[d]	6.4	4.5	3.6	4.5	56	50	76
Eastern Europe	9.4	6.6	6.4	8.0	68	NA	NA
U.S.S.R.	23.0	16.1	10.2	12.8	44	NA	NA
Other centrally planned	8.9	6.2	4.2	5.3'	47	NA	NA
Western world	101.5	71.1	59.0	73.9	58	51	91
Centrally planned[e]	41.3	28.9	20.8	26.1	50	NA	NA
World[f]	142.8	100	79.8	100	56	NA	NA

a. Original data were obtained from British Sulphur Corporation (1982a, 1982b). However, the calculations were made by the authors.
b. Obtained by multiplying the percentage share of sulfuric acid used in fertilizer sector by the percentage share of sulfur used to manufacture sulfuric acid during 1980.
c. Percent fertilizer sulfuric acid used for production of phosphate fertilizers in 1979. Derived from British Sulphur Corporation (1981). Such data were not available for 1980 or 1981.
d. Africa includes South Africa; Asia includes Japan; and Latin America excludes Cuba.
e. Sum of Eastern Europe, U.S.S.R., and Other Centrally Planned Economies.
f. Totals are approximate due to rounding of data.
NA = Not available.

in America and Oceania, has not significantly changed in Asia, and has increased in Africa. The disaggregated analysis by country and region with respect to sulfuric acid consumption and its allocation to the fertilizer industry is given in Table 9.6. Except in a few countries, a major share of sulfuric acid is used in the phosphate fertilizer industry.

A regional summary of S consumption and its estimated approximate allocation among sulfuric acid, fertilizer, and phosphate fertilizer industries are given in Figure 9.1.[3] On the basis of these results and previous discussions, we can conclude that (1) a major share of the S consumed in the

3. In the regional summary reported in Figure 9.1, Africa includes SouthAfrica, Asia includes Japan, and Latin America includes Cuba.

Table 9.6. World consumption and allocation of sulfuric acid and sulfur to fertilizer sector during 1981[a].

Region/country	Total consumption of sulfuric acid '000 mt	Consumption of sulfuric acid by fertilizer sector		Estimated % of sulfur consumption used in fertilizer sector[b]	% used for phosphate fertilizer[c]
		'000 mt	% allocation		
Western Europe	25 192	10 924	43	38	NA
Belgium	2 378	1 128	47	42	66
Finland	1 107	533	48	40	100
France	4 100	2 060	50	45	97
Germany, F.R.	3 700	400	11	10	70
Greece	952	887	93	83	89
Italy	2 455	720	29	25	90
Netherlands	1 794	946	53	NA	100
Spain	3 045	1 864	61	54	72
Sweden	760	245	32	26	100
United Kingdom	2 915	793	27	24	93
Others	1 986	1 348	68	NA	NA
North America	40 820	24 695	60	51	NA
Canada	3 800	2 200	58	47	92
United States	37 020	22 495	61	52	97
Oceania	2 530	1 990	79	73	NA
Australia	1 990	1 450	73	68	98
New Zealand	540	540	100	94	100
Africa	9 009	7 249	83	81	NA
Egypt	245	200	82	NA	100
Morocco	2 305	2 300	100	NA	100
South Africa	3 300	2 200	67	64	93
Tunisia	2 220	2 180	98	96	100
Others	939	369	39	NA	NA
Asia	11 979	5 698	48	43	NA
India	2 700	1 650	61	52	77
Japan	5 914	1 853	31	28	68
South Korea	1 298	898	69	64	100
Taiwan	900	522	58	NA	25
Others	1 167	775	66	NA	NA
Latin America[d]	6 745	4 031	60	54	NA
Argentina	275	31	11	NA	100
Mexico	3 000	2 000	67	62	60
Brazil	2 600	1 750	67	60	100
Chile	385	10	3	NA	NA
Venezuela	130	115	88	NA	40
Others	355	125	35	NA	NA

Table 9.6. Continued.

Region/country	Total consumption of sulfuric acid '000 mt	Consumption of sulfuric acid by fertilizer sector		Estimated % of sulfur consumption used in fertilizer sector[b]	% used for phosphate fertilizer[c]
		'000 mt	% allocation		
Centrally planned[e]					
Eastern Europe	NA	NA	NA	NA	NA
U.S.S.R.	NA	NA	NA	NA	NA
Other centrally planned	NA	NA	NA	NA	NA
Western world	96 275	54 587	57	50	NA
Centrally planned[e]	NA	NA	NA	NA	NA
World[f]	NA	NA	NA	NA	NA

a. Original figures were obtained from British Sulphur Corporation (1982a). However, all the calculations were made by the authors.
b. Obtained by multiplying the percentage share of sulfuric acid used in fertilizer sector during 1981 by the percentage share of S used to manufacture sulfuric acid during 1980 (1981 data not available).
c. Derived from British Sulphur Corporation (1981). The share refers to that part of sulfuric acid in the fertilizer sector which was used to produce phosphate fertilizers.
d. Excluding Cuba.
e. Sum of Eastern Europe, U.S.S.R., and Other Centrally Planned Economies.
f. Totals are approximate due to rounding of data.
NA = Not available.

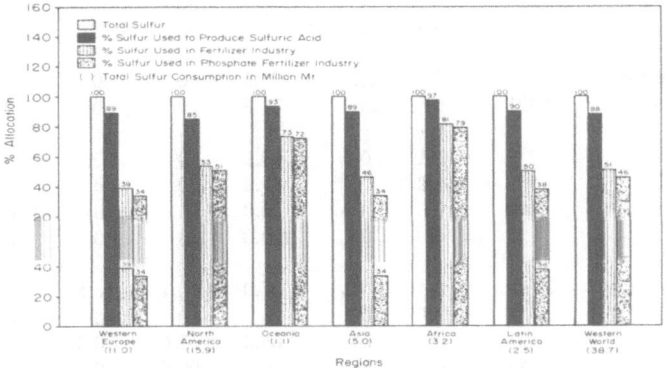

Figure 9.1. Estimated Percent Share of Sulfur Consumption Components in the Western World During 1980.

world is used in the fertilizer industry, especially for phosphate fertilizers; (2) most of the S used in the phosphate fertilizer industry (with the exception of SSP) is discarded as byproduct phosphogypsum; (3) one essential plant nutrient, S, is used to provide another essential plant nutrient, P; and (4) sulfuric acid will continue to play an important role in the phosphate fertilizer industry in the near future.

Patterns and Determinants of World Trade in Sulfur

Sulfur production in the world during 1981 was 53.8 million mt, of which 16.1 million mt (about 30%) was traded in the international market. Sulfur which moves in the international market consists primarily of brimstone. During 1981 brimstone accounted for 96% (15.5 million mt) of all the S trade. The remaining 4% (0.6 million mt) was S in the form of pyrites. As

Table 9.7. Patterns of world trade for brimstone during 1981[a].

Importing region/country	Total imports[b] '000 mt	Exporting countries (sources of imports) share in total imports (%)						
		United States	Mexico	Canada	France	Poland	Others[c]	All[c]
Western Europe	3 957	17.6	6.5	16.2	18.0	32.7	9.0	100
Belgium	392	63.8	0	10.7	12.0	5.6	7.9	100
France	569	11.2	0.9	22.5	0	62.7	2.6	100
Germany, F.R.	339	23.6	0	25.1	0.6	50.4	0	100
Greece	252	5.2	0	11.9	18.3	65.1	0	100
Italy	441	0	0	60.1	18.4	17.2	4.3	100
Netherlands	459	32.9	0	0	21.1	14.4	31.8	100
United Kingdom	856	11.2	14.1	10.7	30.0	33.6	0.4	100
Others	636	6.8	20.6	0	28.6	23.6	20.1	100
North America	2 331	0.4	28.1	71.5	0	0	0	100
United States	2 321	0	28.2	71.8	0	0	0	100
Oceania	776	0.1	0	99.9	0	0	0	100
Australia	569	0.2	0	99.8	0	0	0	100
New Zealand	202	0	0	100.0	0	0	0	100
Others	5	0	0	100.0	0	0	0	100
Africa	2 395	3.4	0.2	67.0	10.1	18.5	0.8	100
Egypt	48	100.0	0	0	0	0	0	100
Morocco	877	0	0	66.0	3.6	30.4	0	100
Niger	53	0	9.4	0	90.6	0	0	100
South Africa	573	2.4	0	97.6	0	0	0	100
Tunisia	770	0	0	57.5	19.5	23.2	0	100
Others	74	25.7	0	31.1	16.2	0	27.0	100

Table 9.7. Continued.

Importing region/country	Total imports[b] '000 mt	Exporting countries (sources of imports) share in total imports (%)						
		United States	Mexico	Canada	France	Poland	Others[c]	All[e]
Asia	1 991	7.7	4.5	59.6	0	2.9	25.2	100
India	924	15.5	9.7	50.0	0	6.2	18.6	100
Indonesia	93	0	0	94.6	0	0	5.4	100
Israel	147	0	0	100.0	0	0	0	100
South Korea	380	1.3	0	53.2	0	0	45.5	100
Taiwan	254	0	0	78.3	0	0	21.7	100
Others	188	2.1	0	48.4	0	0.5	48.9	100
Latin America	1 091	12.0	1.3	66.1	0	12.4	8.2	100
Mexico	50	100.0	0	0	0	0	0	100
Argentina	88	6.8	0	93.2	0	0	0	100
Brazil	801	5.6	1.0	69.2	0	16.9	7.4	100
Chile	100	14.0	0	76.0	0	0	10.0	100
Others	52	28.8	19.2	9.6	0	0	42.3	100
Eastern Europe	1 477	11.4	10.4	0	0	72.1	5.9	100
Czechoslovakia	515	0	0	0	0	88.9	11.1	100
Germany, D.R.	173	0	0	0	0	97.1	2.9	100
Romania	432	35.0	17.4	0	0	44.4	3.2	100
Others	357	5.0	22.4	0	0	69.7	3.4	100
Other centrally planned	1 548	0	1.3	35.0	0	52.9	10.8	100
U.S.S.R.	967	0	2.1	15.9	0	82.0	0	100
China	232	0	0	99.1	0	0	0.9	100
Cuba	315	0	0	50.2	0	0	49.8	100
Others	34	0	0	0	0	76.5	23.5	100
Western world	12 540	8.6	8.1	52.6	7.6	15.4	7.7	100
Centrally planned[d]	3 025	5.6	5.7	17.9	0	62.3	8.4	100
World[e]	15 565	8.0	7.7	45.9	6.1	24.5	7.9	100

a. Original trade figures were obtained from British Sulphur Corporation (1982a). However, all the calculations were made by the authors.
b. World trade in pyrites during 1981 was 585 thousand mt, which is approximately 3.6% of total S (brimstone and pyrites) trade.
c. This includes exports, in '000 mt, from Germany, F.R. (419), Japan (232), U.S.S.R. (203), and others (369).
d. Sum of Eastern Europe and Other Centrally Planned Economies.
e. Totals are approximate due to rounding of data.

174

far as S production is concerned, the respective contributions of brimstone and pyrites to world S production in 1981 were 63% and 21%. Clearly, brimstone is the main form of S in worldwide trade.

The patterns of world trade, including sources, destinations, andquantities traded for brimstone during 1981, are summarized in Table 9.7. Canada and Poland are the major S exporters, accounting for about 70% of all brimstone exports and 68% of all S exports (Figure 9.2). Other important S-exporting countries include the United States (7.7%) from North America, Mexico (7.4%) from Latin America, France (5.9%) and Federal Republic of Germany(2.6%) from Western Europe, Japan (1.4%) from Asia, and U.S.S.R. (2.6%) from centrally planned economies. The emerging S exporters include the Middle Eastern countries. However, given the political situation in the region the potential production and exports of S are subject to uncertainties.

The share of Frasch S in brimstone exports may decline in the future, primarily because of (1) higher energy costs since Frasch mining of S is energy intensive and (2) an increased recovery of S, necessitated by the more stringent SO_2 emission standards in North America and Western Europe. These trends may result in declining importance of Frasch producers and exporters in international S trade. The cost of production will continue to play an important role in determining the export price. The cost of producing recovered S as compared with Frasch S will depend on the extent to which the investment and operational costs of S recovery are charged to the main product.

Although only a few countries play a major role in the S export market, S imports are diffused over a large number of countries. However, among

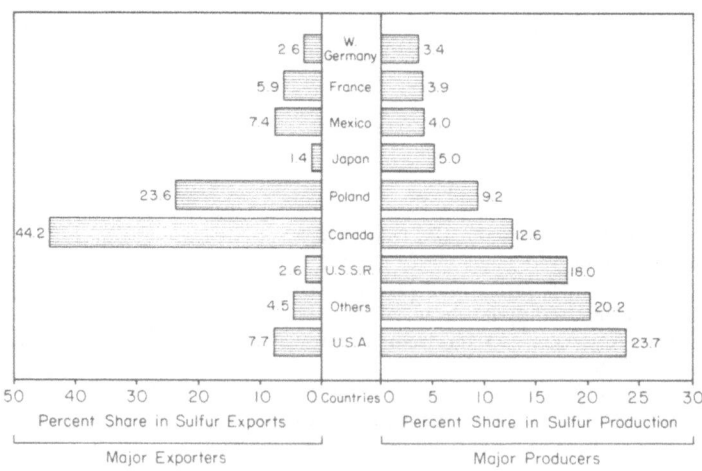

Figure 9.2. Patterns of World Sulfur Production and Exports During 1981.

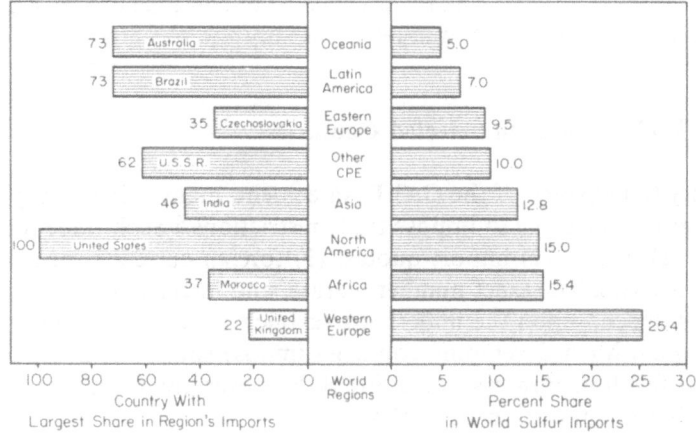

Figure 9.3. Market Share in World Sulfur Imports by Different Regions and Countries During 1981.

individual countries, the United States is the largest producer, consumer, importer, and third largest exporter of S in the world. Other important S-importing countries include the nited Kingdom, Australia, Morocco, Tunisia, India, Brazil, Czechoslovakia, and U.S.S.R. (Figure 9.3). The amount of S imported by each country, of course, depends on the amount domestically produced and total S needs. Three other factors that determine a particular country's financial capability for importing S are the f.o.b. price, shipping arrangements, and the foreign exchange situation. With the exception of Mexico, most developing countries are net S importers. Despite serious foreign exchange scarcity, these countries must import S in order to produce more food and to develop their agricultural and industrial sectors.

The current S trade patterns have evolved over a period of time. Poland exports mainly to Western Europe, Eastern Europe, and other centrally planned countries. During 1981, 62% of the S imports in the centrally planned economies (Eastern Europe and other centrally planned countries) came from Poland. France exports to other West European and Francophone African countries. A major share of Mexico's S exports goes to the United States. The United States exports mainly to Western Europe and Latin American countries. Canada is in a unique position and exports S to almost all the importing countries, except Eastern Europe. The key factors that determine a particular country's relative position in the international S market include (1) quantity produced and cost of production; (2) f.o.b. price and other commercial terms of sale; (3) distance from the importing country and hence the transportation cost; (4) reliability of S supply; and (5) long-term bilateral arrangements between the exporter and the importer.

In view of these criteria, Canada is expected to continue to play a dominant role in the international S market, especially the western world.

Sulfur Trade Policies in Selected Countries

Contrary to the spirit of the GATT Agreement, many S-importing developing countries continue to impose trade restrictions.[4] Such trade restrictions include (1) tariffs, (2) import quotas, (3) foreign currency control, and(4) import licenses. The exact number, nature, and intensity of these trade restrictions depend upon the underlying (explicit or implicit) objective of the government. Most developing countries impose some sort of trade restrictions on fertilizer imports, including S. However, the most common trade barrier is the imposition of tariffs.

Unlike excise duty, which is levied as a fixed amount on each unit sold, tariffs are levied ad valorem. In other words, tariffs are levied at some fixed percentage of the price of commodity. As an example, tariffs on S imports in selected developing countries during 1971 were 10% to 50% ad valorem. The specifics of current tariff policies may have changed since 1971, but the general attitude of developing countries toward import restrictions has not changed significantly. It is primarily a response to the imposition of various trade restrictions by the industrialized countries on the commodities exported by developing countries.

Tariffs on S imports are generally imposed to achieve one or more of the following objectives: (1) to protect the domestic S industry from undue foreign competition (infant industry agreement), (2) to discourage S imports, (3) to encurage the development of S substitutes, and (4) to raise government revenue. Most of the S is imported as brimstone, which is the basic raw material used to manufacture sulfuric acid. Sulfur and sulfuric acid are essential for economic growth, and they do not have satisfactory substitutes. Technically, nitric acid can serve as a possible substitute for sulfuric acid in the phosphate fertilizer industry, but its use has no comparative economic advantage. Besides, the production of nitric acid is highly energy intensive. For countries with limited domestic S resources, tariffs on S imports are convenient sources of government revenue.

Tariffs on S may be a good source of revenue, but they can result in high social costs, especially for a developing country with food deficits and serious soil fertility problems, including S deficiency (Mudahar, 1978). First, the immediate impact of the tariff is to raise the price of imported S by as

4. The GATT Agreement (General Agreement on Tariffs and Trade) was originally signed in 1959. The stated aims are (1) reciprocal and continuing reduction of tariffs and abolishment of other barriers of trade and (2) nondiscrimination in commercial trade. In addition to GATT, there are many regional and multilateral organizations designed to achieve similar aims among member countries.

much as the tariff. Second, the tariffs encourage inefficiency in the domestic S industry and in those industries that use S or S derivatives. Third, in the absence of any subsidy, the incidence of tariffs on S imports is passed on to consumers and farmers who end up paying higher prices. The net economic impact of a tariff depends on many factors. However, for a low-income country, S tariffs may not be in the best interest of the general population since they encourage both inefficiency (higher cost of fertilizer production) and inequity (transfer of income from fertilizer users to fertilizer producers).

Behavior of International Sulfur Prices

For the majority of the S-importing, low-income countries, international S prices exert an important influence on national decisions related to S imports, S consumption, and the contribution of S to economic growth.

The international S prices are determined by several factors some of which may be unique to the S market. First, the international S market is dominated by brimstone as the primary product that moves in the world market. Second, the Frasch S producers (mainly Canada and the United States) generally maintain large stocks of S as an essential part of their marketing strategy in order to ensure long-range supply reliability and to provide a cushion for shortterm fluctuations in price. Third, the S market can be broadly characterized as an oligopoly (i.e., a market in which there are a few large S suppliers, a relatively homogeneous product, and barriers to entry). The behavior of the S market tends to be monopolistic under tight market conditions and competitive under surplus S supply conditions. Fourth, S demand can be characterized as a 'derived' demand that is greatly influenced by demand for primary products. In the absence of artificial restraints, the international S market in the short run can be characterized by relatively inelastic demand and elastic supply.

Prior to the 1950s, the United States played an important role in the international S market. According to Hazleton (1970), the export prices for Frasch S were determined by the Sulphur Export Corporation (Sulexco), throuh a cartel agreement with the Sicilian producers. The export prices exceeded the domestic S prices and were much higher than the level of marginal cost of production. The rate of return on average invested capital was abnormally high, averaging 23.60% from 1919 to 1953. However, during periods of S glut, the suppliers intervened in the S market by offering hidden incentives (i.e., freight absorption, price discounts) without changing the posted S prices. The international S market became more competitive after 1955, with the entry of Frasch S from Mexico and recovered S from Canada and France. This, however, does not mean that the monopolistic elements of the international S market have disappeared-they are merely dormant.

178

The behavior of the international S market and prices are reported in Figure 9.4 by annual price levels and changes.[5] This diagram is based on S prices in the United States, which generally reflect levels and trends in international S prices. From the prices reported in the diagram, we can draw the following conclusions.

First, S prices expressed in constant dollar terms generally declined from 1955 to 1978 (with the exceptions of two price upswings peaking in 1968 and 1975). The decline in real prices of Frasch S was due to the decline in unit cost of production resulting from the decline in energy prices and economies of scale in mining.

Second, as indicated by current prices, the international S market has not been as stable as many experts claim. With the exception of the period from 1955 to 1963, average annual prices have indeed been fluctuating a great deal.

Third, S prices in the export market have generally been higher than the prices in the domestic market. A large part of this gap may be explained by the transportation costs. However, as has been indicated by Hazleton

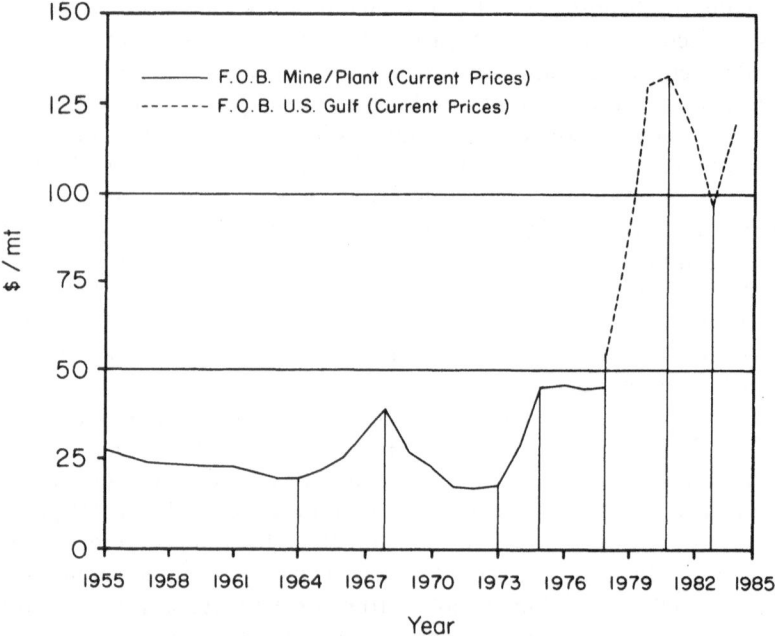

Figure 9.4. The Dynamics of Sulfur Prices in the United States (Average Annual Price).

5. The figure is based on recent prices obtained from different sources. The f.o.b. mine/plant prices in the United States are obtained from Shelton (1979). On the other hand, f.o.b. U.S. Gulf prices are derived from prices reported in different issues of Green Markets (1984).

(1970), price discrimination between the domestic and export market was always there prior to 1960 and that can be partly explained by the cartel agreements between Sulexco and major foreign producers of elemental S.

Fourth, S prices can be characterized as −administered− prices in the sense that prices are determined by administrative decisions that partly reflect the S market conditions. For example, f.o.b. (U.S. Gulf) prices for dry S remained constant at $132.50/mt for 12 months from March 1981 to February 1982.

The historical behavior of S prices, as reported in Figure 9.4, can be divided into the following eight phases:

1. 1955-64:   Stable and declining.
2. 1964-68:   Sudden increase, peaking in 1968.
3. 1968-73:   Equally dramatic drop.
4. 1973-75:   Sellers' market.
5. 1975-78:   Almost constant and stable.
6. 1978-81:   Tight market, skyrocketing prices.
7. 1981-83:   Very high, but declining.
8. 1983-To date: Very high and rising rapidly.

These price fluctuations are the result of many factors. First, since a large share of S is consumed in the industrialized countries, the general economic situations in these countries play an important role in determining S demand and hence S prices. Second, the developments in the fertilizer industry, in terms of demand and price for fertilizers using S, exert an important influence on S demand and prices. Third, an increase in energy prices results in higher cost of production for Frasch S. For example, an increase in energy prices in 1974/75 and 1979/80 was reflected in higher S prices. Fourth, because S is a bulky commodity, the logistical problems of moving S from suppliers to meet sudden spurts in demand create temporary supply shortages. Sulfur prices, in fact, declined from a high of $135.00/mt in January 1981 to a low of $88.00/mt in October 1983; since then, however,

Table 9.8. Sulfur freight rates and their contribution to international sulfur prices[a].

Year	Dry sulfur price, f.o.b. Canada ($/mt)	Freight rate for sulfur in bulk, f.i.o. basis from Vancouver to India ($/mt)	Freight as % of f.o.b. sulfur price
1977, May	41.50	22.00	53
1978, May	41.00	20.90	51
1979, February	56.00	27.25	49
1980, April	126.30	59.00	47
1981, June	127.50	53.00	42
1982, June	110.00	30.00	27

a. Derived from different issues of British Sulphur Corporation (1983b).

the prices started rising again to a new high of $145.90/mt in December 1984.

In addition to the f.o.b. price, freight rates play an important role in determining c.i.f. prices of S (Table 9.8). The freight rate as a percentage of the f.o.b. price of S for India declined from 53% in May 1977 to 27% in June 1982. Despite a sharp decline, however, the freight rate in the absolute sense is still quite high. Both S prices and freight rates have dropped recently, but the relative drop in S prices is smaller than in freight rates. The S-importing developing countries need to carefully evaluate not only the f.o.b. prices but also the freight rates since it is the c.i.f. price that matters to them, especially when foreign carriers are involved.

Sulfur Reserves, Resources, and Their Use

Sulfur is known to occur in many different forms and deposits. It is considered one of the more abundant elements on earth, ranking 13th in magnitude. It has been estimated that S accounts for 0.052% of the earth's crust. However, this does not mean that S in commercial forms is available everywhere in abundance. Only a small fraction of the large S resources is sufficiently concentrated to make the recovery feasible at the current state of knowledge, technology, and prices.

Depending on the precise knowledge about the location, quantity, chemical composition, and cost of recovery, S resources could be considered as (1) identified, (2) probable, and (3) speculative. Sulfur reserves are only a small subset of S resources. Sulfur reserves include only that part of S resources that are known and can be recovered profitably at current knowledge, technology, and prices. In response to positive changes in any or all of these variables, both the amount of S reserves and different forms of S resources increase. The S resources are already there: either they become known through better exploration techniques, or they become profitably recoverable through better technology and higher prices.

The known S resources can be broadly classified into the following 11 categories:
1. Evaporites.
2. Volcanic rocks.
3. Natural gas.
4. Petroleum.
5. Pyrites.
6. Metallic sulfides.
7. Tar sands.
8. Coal.
9. Oil shale.
10. Gypsum/anhydrite.
11. Seawater.

Table 9.9. Estimated sulfur resources in the world[a]. (million mt)

Type of resource	Identified	Probable	Total
1. Evaporites	580	>250	>830
2. Volcanic rocks	130	>40	>170
3. Natural gas	170	885	1 055
4. Petroleum	265	1 330	1 595
5. Pyrites	650	>40	>690
6. Metallic sulfides	360	>440	>800
Subtotal (A)	2 155	>2 985	>5 140
7. Tar sands	50	>1 800	>1 850
8. Coal	33 450	199 600	233 050
9. Oil shale	–	281 000	>281 000
10. Gypsum	>7 200	Vast	Vast
11. Seawater	–	Unlimited	Unlimited
Subtotal (B)	>40 700	>482 400	>523 100
Total	42 855	>485 385	>528 240

a. Derived from Bixby (1979, 1980), Horseman (1973), Meyer (1977), and Shelton (1979).

The amount of S contained in each of these resources varies with the location, type of deposit, and chemical composition.[6]

The estimated S resources in the world are reported in Table 9.9. According to these estimates, total S resources are more than 500 billion mt, of which approximately 10% is considered identified. The share of S reserves is even smaller, less than 1% of the total S resources. As far as the type of S resources is concerned, about 98% of known S is contained in coal and oil shale.[7] These estimates do not include vast S resources contained in anhydrite, gypsum, and seawater. The location and magnitude of known S reserves are given in Table 9.10. The amount of S reserves in the world varies between 1.8-2.2 billion mt, rather a small fraction of total S resources. Of these reserves, the developed countries account for approximately 65%. About 29% of S reserves are located in China (1.4%), Mexico (5.1%), Iraq

6. The S content of some of these minerals in pure forms is 53.4% S in pyrites, 18.6% S in natural gypsum, and 23.5% S in anhydrite. The sulfate content in seawater is estimated at 2,760 ppm, whereas total dissolved salt content is 36,000 ppm. Furthermore, according to Meyer (1977), S content by weight in selected hydrocarbon is 0.05%-14% in crude oil, 4% in tar sand bitumen, 1% in shale oil, and 1%-14% in dry bituminous coal.

7. According to the President's Commission on Coal (1980), the estimated share of recoverable coal reserves in the world (650 billion mt) is located as follows: 31% in the United States, 23% in U.S.S.R., 21% in Europe, 14% in China, 4% in Oceania, 1% in Canada, and the remaining 6% in the rest of Asia (2.9%), Africa(2.6%), and Latin America (0.5%).

182

(8.5%), and remaining countries in the Near East (14.2%).[8] The rest of the developing tropical countries in Asia, Africa, and Latin America account for less than 6% of the known S reserves.

As reported in Table 9.11, these low-income tropical countries are known to have S resources in the form of anhydrites, gypsum, and nonferrous sulfides. Some of them are even known to have elemental S deposits and pyrites. The important questions that need to be answered are concerned with(1) the agronomic effectiveness and techno-economic feasibility

Table 9.10. Location and magnitude of identified world sulfur reserves[a].

Region/country	Sulfur reserves[b]	
	Million mt	% share
North America	425	24.1
United States	175	9.9
Canada	250	14.2
Latin America	125	7.1
Mexico	90	5.1
Others	35	2.0
Europe	690	39.1
U.S.S.R.	250	14.2
Poland	150	8.5
France	30	1.7
Germany, F.R.	30	1.7
Spain	30	1.7
Italy	15	0.8
Others	185	10.5
Asia	485	27.5
Japan	10	0.6
Iraq	150	8.5
China	25	1.4
Others[c]	300	17.0
Africa	20	1.1
Oceania	20	1.1
World	1 765	100

a. Derived from Shelton (1979).
b. These reserves are assumed to include sulfur mainly from evaporites, volcanic rocks, natural gas, petroleum, pyrites, and metallic sulfides.
c. About 250 million mt is estimated to be in Near East, excluding Iraq.

8. Figures in parentheses refer to the estimated percentage share of the individual country or region in question.

of these S resources and (2) how the indigenous S resources can be used to meet the S requirements of countries in Asia, Africa, and Latin America. Any positive technical assistance, financial aid, and research efforts directed toward these questions can result in a large economic contribution to the developing tropical countries through alleviation of their food and foreign exchange problems.

Is there an S resource problem? It depends on how one looks at the problem, i.e., from a physical point of view or from an economic point of view. The physical view is straightforward and addresses the question, How long will it take before we run out of known S reserves if we continue to use S at a postulated growth rate? The economic view, on the other hand, accounts for such variables as recovery costs, sale price, profitability, and economic rent. From this point of view, the S resources may be there, but their recovery may not be economical. The economic view is greatly influenced by technological change and government policy. From a physical point of view, the life expectancy of world S resources is 331 years at 5% annual growth in production from 1972/74 onward (Tilton, 1977). The new S reserves are constantly increasing over time. For example, S reserves in 1974 were five times more than in 1950. Furthermore, given current knowledge about the amount of reserves and resources, current level of depletion, and a postulated growth in depletion, we may run out of S reserves and resources before we run out of either phosphate or potash resources. However, these conclusions may need to be modified once we account for the impact of other variables such as economics, technical change, and government policy.

Another important question that we need to consider is what would be the S recovery cost, S prices, and their impact on fertilizer prices by the year 2000? It is very important here to make a clear distinction between the private costs and social costs. Many factors would influence these costs, future S prices, and their relative impact on fertilizer prices. Some of these factors are as follows: (1) quality of new S resources; (2) technological change with respect to exploration techniques, mining, and recovery processes; (3) economies of scale; (4) energy prices and availability; (5) capital costs and its availability; (6) logistics; (7) environmental concern; (8) impact of mining on land use; (9) government regulations; and finally (10) discovery of S substitutes. The respective governments can play a crucial role in modifying any of these variables with subsequent implications for S demand, supply, and prices. This includes government policy with respect to (1) education and training, (2) research and development, (3) economic incentives, and (4) protection of environment and health of citizens.

184

Table 9.11. Type and location of known sulfur resources in individual countries of the world[a].

Region/country	Type of sulfur resources[b]					
	Elemental	Natural gas	Petroleum	Pyrites	Metallic sulfides	Gypsum
Western Europe						
Austria	–	–	R	–	C	A,G
Belgium	–	–	R	–	Z	–
Cyprus	–	–	–	P	C	G
Finland	–	–	–	P	C	–
France	N	NG	R	P	–	–
Germany, F.R.	–	NG	R	P	C,Z	–
Greece	V	–	–	P	C,L,Z	–
Italy	N	NG	R	P	C	–
Norway	–	–	R	P	NS	–
Portugal	–	–	R	P	NS	–
Spain	–	–	–	P	NS	–
Sweden	–	–	–	P	NS	–
Turkey	N	–	R	P	NS	G
United Kingdom	–	–	R	–	–	A,G
Eastern Europe						
Bulgaria	–	–	R	–	NS	–
Czechoslovakia	–	–	R	–	NS	–
Germany, D.R.	–	–	–	–	NS	A
Poland	N	–	R	P	NS	–
Romania	–	–	R	–	NS	–
U.S.S.R.	N	NG	R	P	NS	–
Yugoslavia	–	–	–	–	NS	–
North America						
Canada	N	NG	R	P	NS	–
United States	N	NG	R	P	NS	–
Oceania						
Australia	–	–	–	P	NS	–
New Zealand	N	–	–	P	NS	–
Latin America						
Argentina	N	–	R	–	NS	–
Bolivia	N	–	–	–	–	–
Brazil	–	–	R	P	–	–
Chile	V	–	–	–	NS	–
Colombia	V	–	R	–	–	–
Mexico	N	NG	R	–	–	–
Peru	V	–	–	–	NS	–
Venezuela	V	NG	R	–	–	–

Table 9.11. Continued.

Region/country	Type of sulfur resources[b]					
	Elemental	Natural gas	Petroleum	Pyrites	Metallic sulfides	Gypsum
Africa						
Algeria	–	NG	–	–	NS	A
Angola	N	–	–	P	–	A
Benin	–	–	–	–	–	G
Egypt	N	–	R	–	–	A,G
Ethiopia	N	–	–	–	–	A,G
Kenya	–	–	–	–	–	G
Libya	–	NG	R	–	–	A,G
Malawi	–	–	–	P	NS	G
Mali	–	–	–	–	–	G
Morocco	–	–	–	P	NS	G
Mauritania	–	–	–	–	–	G
Mozambique	–	–	–	P	NS	–
Niger	–	–	–	–	–	G
Somalia	N	–	–	–	–	G
South Africa	–	–	–	P	NS	G
Sudan	–	–	–	–	NS	G
Tanzania	–	–	–	–	NS	A,G
Tunisia	–	–	–	–	NS	–
Uganda	–	–	–	–	NS	G
Zaire	–	–	–	–	NS	G
Zambia	–	–	–	–	NS	G
Zimbabwe	–	–	R	P	NS	–
Asia						
Afghanisthan	N	NG	–	–	–	G
Burma	–	–	–	–	NS	–
China	N	–	R	P	–	–
India	–	–	R	P	NS	A,G
Indonesia	N	–	R	–	–	–
Iran	N	NG	R	–	–	–
Iraq	N	–	R	–	–	–
Japan	V	–	R	P	NS	–
Kuwait	–	–	R	–	–	–
Pakistan	N	–	–	–	–	A,G
Philippines	V	–	–	–	NS	–
Saudi Arabia	–	–	R	–	NS	–
Thailand	–	–	–	P	–	A,G

a. Derived primarily from British Sulphur Corporation (1974). However, other available sources were also used to develop this table. For example, International Petroleum Encyclopedia (1982) was used to check natural gas and petroleum reserves.

b. – indicates no known deposits. With further exploration and by improved exploration techniques, these countries may find sulfur resources that are not currently known. The abbreviations used in the table refer to the following: N = native evaporites; V = volcanic rocks; NG = natural gas; R = crude oil refining based on domestic or imported crude; P = pyrites; NS = nonspecified nonferrous sulfides; C = copper sulfides; Z = zinc sulfides; L = lead sulfides; A = anhydrite; and G = gypsum.

Production and Use of Phosphogypsum

Phosphogypsum is a byproduct of wet-process phosphoric acid, a process in which phosphate rock is reacted with the sulfuric acid. Phosphogypsum isgenerally considered a waste product, even though all the S from the sulfuric acid is contained in phosphogypsum. Depending on the location of the wet-process phosphoric acid plant, phosphogypsum is either discarded into the ocean or river, or stored in ponds or heaps. For those countries that (1) import S, (2) have no known S resources, and (3) have widespread S deficiency, phosphogypsum provides a potentially economic (accounting for the opportunity cost of foreign exchange) source of S for the agricultural sector. On the other hand, if all the phosphogypsum is left unused, it could result in serious storage and environmental pollution problems.

The economic importance of phosphogypsum cannot be assessed in the absence of reliable quantitative estimates of its production in selected world regions and developing countries. The following simple mathematical model was used to derive quantitative estimates for phosphogypsum production and S contained in phosphogypsum:

1. $PGP_{jt} = PC_{jt} * CUR * PGP/T,$

2. $SPG_{jt} = PGP_{jt} * S/TPG,$

3. $CPGP_j = \sum_{t=1967}^{1985} PGP_{jt}$

4. $CSPG_j = \sum_{t=1967}^{1985} SPG_{jt},$ or

5. $CSPG_j = CPGP_j * S/TPG .$

Where

PGP_{jt} → Phosphogypsum production in country/region j and year t;

PC_{jt} → P_2O_5 capacity in country/region j and year t;

CUR → Average capacity utilization rate;

PGP/T → Phosphogypsum production per metric ton of P_2O_5;

SPG_{jt} → Sulfur contained in phosphogypsum in country/region j and year t;

S/TPG → Sulfur contained in 1 mt of phosphogypsum;

$CPGP_j$ → Cumulative production of phosphogypsum from 1967 to 1985; and

$CSPG_j$ → Cumulative amount of sulfur contained in phosphogypsum from 1967 to 1985.

187

On the average, the capacity utilization rate of wet-process phosphoric acid plants was assumed to be 80% which implies that CUR : 0.8. Since the capacity utilization rate is plant specific and varies over time, the phosphogypsum production estimates should be considered only as good first-order approximations (which result in overestimates for some developing countries and underestimates for industrialized countries). In the absence of impurities, 4.62 mt of phosphogypsum is produced for every 1 mt of P_2O_5. Phosphate rock is rarely free of impurities, however. In this respect, phosphogypsum production will be underestimated. Finally, phosphogypsum is assumed to contain 17% S, on the average, which implies that S/TPG : 0.17.

The annual (for 1982) and cumulative (from 1967 to 1985) estimates of phosphogypsum production and S contained in phosphogypsum are reported in Table 9.12. During 1982, the estimated production of phosphogypsum

Table 9.12. Estimated production of phosphogypsum and sulfur contained in phosphogypsum in selected countries, regions, and the world[a]. (million mt)

Countries/regions	Phosphogypsum		Sulfur in phosphogypsum	
	Production in 1982	Cumulative production from 1967 to 1985	During 1982	Cumulative from 1967 to 1985
India	2.5	28	0.42	4.7
Philippines	0.3	6	0.05	1.0
Zimbabwe	0.1	2	0.01	0.3
Brazil	2.4	21	0.42	3.6
Mexico	2.0	37	0.34	6.2
North America	39.8	600	6.77	101.9
Western Europe	16.3	279	2.76	47.4
Eastern Europe	8.3	99	1.40	16.9
U.S.S.R.	19.0	201	3.23	34.1
Centrally planned Asia[b]	0.1	2	0.02	0.3
Oceania	1.3	17	0.22	2.9
Far East, DgME[c]	4.6	57	0.78	9.7
Africa, DgME[c]	9.5	95	1.61	16.2
Latin America	4.5	63	0.77	10.7
Others[d]	12.0	144	2.04	24.4
World[e]	115.3	1 556	19.61	264.6

a. Derived from fertilizer capacity data (as of November 1982), originally collected by the National Fertilizer Development Center (NFDC, 1982).
b. Includes China, North Korea, Vietnam, and Mongolia.
c. Developing market economies in Far East and Africa.
d. Includes Near East, Japan, Taiwan, and South Africa.
e. Totals are approximate due to rounding of data.

in the world was 115 million mt, with approximately 20 million mt of S contained in it. The cumulative world production from 1967 to 1985 is estimated to be 1.6 billion mt of phosphogypsum containing 265 million mt of S. One can imagine the gravity of the situation in the year 2000. The developing tropical countries of the Far East, Africa, and Latin America together produced 18.6 million mt of phosphogypsum in 1982, with over 3 million mt of S contained in it. These countries cannot afford to waste such an economic resource, especially when their soils are becoming seriously deficient in S.

According to Weterings (1982), 92 million mt of gypsum (natural plus chemical) was consumed in the whole world during 1981. Of this, only 15 million mt (16%) was chemical gypsum, including 10.65 million mt of phosphogypsum. Most of this phosphogypsum was consumed in Japan, Europe, and the U.S.S.R. A logical question follows: Why cannot phosphogypsum be used in place of natural gypsum? There are several problems to be overcome before phosphogypsum can become a substitute for natural gypsum in current uses. According to Weterings (1982), the current uses of gypsum or phosphogypsum are as follows:

Use	Phosphogypsum Consumption in Western Europe % Share in 1981	Gypsum Consumption in World, % Share in 1981
Building products	57	61
Setting retarder for cement	23	24
Production of ammonium sulfate	8	4 (approx.)
Sulfuric acid and cement	8	1 (approx.)
Miscellaneous	4	10 (approx.)
Total	100	100
	(2.6 million mt)	(92 million mt)

The phosphogypsum cannot be readily substituted for natural gypsum in current uses because of technological and economic problems.

Phosphogypsum is highly contaminated with phosphoric acid and heavy metal impurities that come from the phosphate rock; it also has a slight amount of radioactivity. These three problems render phosphogypsum technically unsuitable for use in the building industry. The drying and purification costs make the use of phosphogypsum uneconomical as compared with natural gypsum, especially in those countries where natural gypsum is rather cheap and readily available.

According to Agarwal (1982) the use of phosphogypsum as a raw material (as an alternative to S) for manufacturing cement and sulfuric acid is economical, especially when cement and S prices are relatively high. In some European countries, phosphogypsum has been used as a raw material for manufacturing AS. As has been indicated by Frederick (1983) and Weterings (1982), contamination problems pose a threat mainly when phosphogypsum is used in the building industry. These problems may not be serious at all when phosphogypsum is used as a source of fertilizer S in agriculture. This is where the main challenge and promise lie.

Research needs to be intitated in order to determine (1) the agronomic effectiveness of phosphogypsum as a source of S; (2) the technical problems related to drying, transportation, storage, handling, and conversion of phosphogypsum into AS or sulfuric acid; and (3) the economic viability of phosphogypsum as a raw material source for manufacturing AS and sulfuric acid, and direct application of phosphogypsum to the field as a source of fertilizer S.

10 Summary, Conclusions, and Recommendations

Sulfur — A Neglected Fertilizer Nutrient

Sulfur is one of the major plant nutrients. It rivals phosphorus in its uptake by plants and nitrogen in protein synthesis, and it is indispensable for certain essential amino acids. Yet its significance as a fertilizer nutrient has not been recognized, particularly in tropical agriculture.

There are two primary reasons why sulfur has not received adequate attention: (1) low-yield subsistence agriculture has exploited the natural reserves of sulfur in the soil and (2) sulfur has been supplied to agroecosystems from the atmosphere through rain and dust, through gaseous absorption, and through the use of irrigation water, manures, and fertilizers like ammonium sulfate and single superphosphate, which added so much sulfur to the soil that the need for sulfur fertilizers was not felt in many countries.

However, the situation has changed in the last three decades. First, ammonium sulfate and single superphosphate have been replaced with urea and triple superphosphate, respectively, which contain very little sulfur. Second, subsistence agriculture is being transformed through the use of high-yielding crop varieties, greater use of fertilizers, and intensive cropping patterns. These changes are creating a large gap between the sulfur supply and sulfur requirements in the soil system. Thus, the potential of the modern agricultural system is not being fully realized.

The primary objective of this study is to analyze the economic importance of sulfur in the fertilizer industry, food production, and the agricultural sector in the tropical countries. The study is the result of growing awareness of the significance of sulfur and of continuing efforts by IFDC to develop fertilizer technology, improve soil fertility, and identify those public policies that will facilitate growth in agricultural production in the developing tropical countries.

Tropics and Food Production — The Target Area of Study

The central theme of the study is the relationship between fertilizer sulfur and food production in the developing countries of the tropics. The greatest problem of these regions is food (both quality and quantity), and the most serious threat to humanity is hunger and malnutrition caused by the widening gap between the demand for food and its production in the tropics.

The study was restricted to the tropics (humid, subhumid, and semiarid tropics), which is a region covering about 4.96 billion ha of land. Approximately 95% of it lies in Asia, Africa, and Latin America, and it includes

191

areas like India, Indonesia, Bangladesh, Nigeria, Brazil, and Mexico which have some of the world's large populations. Many experts believe that the future of humanity lies in the tropics.

The common man's concept of the tropics is limited to humid and sub-humid tropics; yet the semiarid regions are also an important agricultural part of the tropics. Moreover, most of the sulfur deficiencies that have been reported in literature in the last 30 years are from these regions. The harsh environments and large area of sandy or coarse-textured soils (nearly 452 million ha) make these regions an important component of the target area of the study because their soils are inherently low in organic matter and are highly susceptible to leaching of sulfate sulfur. Sorghum, millet, groundnuts, pulses, soybeans, oilseed, and cotton are the most common crops of the semiarid tropics, and they all have high sulfur requirements.

Sulfur in Plant Nutrition

The role of sulfur as a necessary nutrient for plant growth is undisputed; less clear, however, is the way in which sulfur performs its valuable functions and how it interacts with other nutrients and chemicals in the soil and the living matter. Although considerable empirical information is available, more precise information is needed about these interactions. This is particularly important for tropical countries that consider phosphate and lime the key factors of sound fertilizer practice but fail to appreciate their effect in causing leaching losses of sulfate sulfur.

The disproportionately higher use of nitrogen and phosphate in comparison with sulfur, which is evident from the examples of fertilizer consumption in India and Brazil (Table 10.1), may adversely affect the availability of sulfur to plants. The widening ratios of N:S in fertilizers will lower efficiency of nitrogen utilization because of sulfur deficiency; the widening ratio of P:S will worsen the situation by aggravating sulfur deficiency through loss of sulfur in leaching.

Sulfur also differs from nitrogen in that it is not transferred from old

Table 10.1. Changes in estimated nitrogen:sulfur and phosphate:sulfur ratios in total fertilizer consumption in India and Brazil over time.

Year	N:S ratios		P_2O_5:S ratios	
	India	Brazil	India	Brazil
1960/61	0.9	0.9	0.2	1.1
1970/71	3.8	1.1	0.9	1.5
1980/81	14.1	1.8	4.3	4.1

leaves to growing parts or young leaves; nor can it be fixed as nitrogen can be fixed biologically from the atmosphere. There is a need for collection of sulfur-uptake data for different crops and especially their high-yielding varieties in different tropical regions to develop a better understanding of the problems associated with the use of fertilizer sulfur in crop production.

Sulfur in Food Production and Human Nutrition

A survey of available evidence indicates that sulfur deficiency in soil adversely affects not only crop yields but also the nutritional quality of the crop. The data, although scanty, cannot be overlooked because of the serious nutritional consequences of sulfur deficiency.

Some examples of crops and areas in which sulfur deficiency in the soil has affected nutritional quality are as follows: *Asia* − rice in Sulawesi province of Indonesia; wheat, oilseeds (groundnuts, rape and mustard, soybeas), pulses and potatoes in India; *Latin America* − soybeans, maize, beans, rice, and pasture legumes in Brazil; and *Africa* − millets in Uganda. These examples are warning signals of potentially serious problems for human nutrition.

Sulfur deficiencies in tropical countries cause a reduction in the amount of methionine, cysteine, and cystine types of sulfur-containing essential amino acids in groundnuts, pulses, and cereals that will be disastrous for cereal-consuming countries. The gravity of the problem is intensified by the decline in production of these food commodities and the deterioration in their quality because of sulfur deficiency. The shortage of oilseeds and pulses, the widening protein gap, and increasing malnutrition are well-recognized problems of the developing countries in the tropics, and sulfur deficiency is worsening the situation.

There is much evidence that sulfur fertilization improves the quality of pasture legumes and grasses in all the tropical countries and, thus, directly affects animal health. Consequently, sulfur deficiency affects the quality of food for both human beings and animals in the tropical countries. The magnitude of the problem cannot be quantified accurately because of inadequate research data. However, by using the average amount of sulfur removed by the crops and the appropriate growth rate for the production of each crop, we have estimated the total sulfur removal and hence the likely gap between sulfur supply and requirements by the year 2000.

The general decline in the percentage share of the total sulfur uptake required for the likely production of pulses, oilseeds, and groundnuts in India and Nigeria during 1960, 1980, and 2000 indicates the potential impact of sulfur deficiency on the nutrition of the people who depend on these foods as sourcesof sulfur-bearing amino acids (Table 10.2). The phenomenal rise in production and exportation of soybeans from Brazil is overstraining the

Table 10.2. Estimated proportion of sulfur uptake by pulses and oilseeds in India and Nigeria from 1960 to 2000.

	Percentage of estimated total S uptake by 15 crops/crop groups[a]					
	India			Nigeria		
	1960	1980	2000	1960	1980	2000
Pulses	14.58	8.26	4.86	–	–	–
Oilseeds[b]	4.76	4.57	3.11	0.50	1.33	1.36
Groundnuts	5.29	4.64	3.40	11.68	5.12	3.45

a. Crops and crop groups included in this analysis of the estimated sulfur requirements are wheat, rice, maize, sorghum, millet, pulses, root crops, oilseeds, cotton, groundnuts, sugarcane, tobacco, coffee, soybeans, and oil palm.
b. Other than soybeans and groundnuts.

sulfur reserves of the soil and creating a greater need for their replenishment. The slow growth or even a decline in production of pulses and groundnuts is also an indication that less sulfur is available in the food system in many tropical countries.

Sulfur Status in the Tropics – Additions

The supply of sulfur for plant nutrition depends on (1) the sulfur-supplying capacity of the soil and (2) the addition of sulfur from external sources such as atmosphere, irrigation water, manures, and crop residues, as well as additions from such chemical sources as sulfur-containing fertilizers and pesticides.

Sulfur-Supplying Capacity of Tropical Soils

There is little information about the amounts, forms, and distribution of sulfur in the soil, its availability to crops, and the rate of its disappearance from the agricultural system. However, available information suggests that, in general, the tropical soils of Asia, Africa, and Latin America have low total reserves of sulfur because of low quantities of organic matter and the rapid mineralization as well as leaching losses.

There are some soils with high total sulfur and high organic sulfur but limited available sulfur. The volcanic ash soils, the Andepts, may have a high organic matter content and a large amount of organic sulfur, but they are poor in available sulfur because of the high adsorption and immobilization of sulfur. Hence, they respond to sulfur application.

There is another large group of soils that have low sulfur adsorption capacity in the surface soil but higher sulfur adsorption capacity in the sub-soil; they show a high amount of total sulfur reserves. Yet even in these soils the shallow-rooted crops generally suffer from sulfur deficiency. Deep-rooted crops like cotton may also initially experience a setback because of the low supply of available sulfur; after initial nutrient stress, however, many deep-rooted crops may be able to partially exploit the adsorbed sulfur.

Another group of soils, the coarse-textured soils, have low reserves of sulfur, low sulfur adsorption capacity, and high susceptibility to leaching losses. In such soils, besides determining the amounts of sulfur fertilizers to be used, the main problem is to reduce the loss of applied sulfur. These soils are very responsive to sulfur application.

Notwithstanding the limitations of the methods used for determining available sulfur and the inadequacy of research results from many countries, it is evident that a significant percentage of the soils in tropical regions are deficient in available sulfur. In some cases sulfur deficiency closely followsphosphorus deficiency.

As a result of leaching and lack of replenishment of the nutrients lost, soils of the tropical region, especially the coarse-textured soils and highly weathered soils such as Ultisols, Oxisols, Alfisols, and Inceptisols, are either inherently deficient in sulfur or are likely to become deficient after clearing of the land, burning of the vegetation, and continuous cropping. Sulfur trends in virgin and continuously cropped lands of Brazil indicate that induced sulfur deficiency will soon become a limiting factor for crop production. Clearly any future strategy for increasing food production in new tropical areas must include sulfur fertilizers in the research and development programs.

Furthermore, there is evidence that sulfur deficiency in tropical soils is also aggravated by liming, phosphating, and an imbalanced use of NPK fertilizers that exclude sulfur. Thus, to avoid compounding the adverse effects of the 'Green Revolution,' which is synonymous with the use of high-yielding varieties of cereals and large amounts of nitrogen fertilizers, the use of sulfur-supplying fertilizers becomes necessary for such situations.

Contribution of Sulfur From External Sources

Contribution From Atmosphere — Studies from Nigeria and Kenya show that approximately 2-3 kg S/ha is added annually from the atmosphere to the soil and the amount increases with the rainfall. Such information is not available from the other tropical countries; however, on the basis of evidence from rural areas of tropical regions of Australia, central Kenya, and northern Nigeria, one could not expect the sulfur contribution from the atmosphere to be any highr than these estimates indicate. Nevertheless, additional data need to be collected in other developing countries of the tropics.

Contribution From Irrigation Water — The sulfate in irrigation water could become an important source of sulfur to crops in irrigated areas. However, little research has been done on irrigation as a source of sulfur in most of the tropical countries, particularly in the semiarid tropics where irrigation is becoming an important part of the strategy for increasing crop production.

Irrigation water of satisfactory quality with respect to the salinity and with enough sulfate sulfur could partially meet the sulfur needs of the crops. However, in spite of many normal waters with high sulfate sulfur content in India and elsewhere, sulfur deficiencies have still been observed.

Sulfur deficiency has also manifested itself in areas irrigated with waters of low sulfate content. Thus, the role of irrigation water in contributing sulfur to soil cannot be correctly assessed without more in-depth studies. Some of the conflicting results being obtained from different areas could be attributed to the variable sulfur content of water and the nature of soil.

Contribution by Crop Residues, Manures, and Fertilizers — In the developing countries of the tropics, crop residues usually are either removed or burned. Thus, the addition of sulfur through crop residues is very small. Likewise the average use of fertilizers and manures is so low that the contribution from this source is too small to be of major consequence.

Thus, with the traditional system of farming and subsistence agriculture, the estimated annual additions of sulfur through all the external sources in tropical agriculture for all practical purposes will be no more than 4-5 kg/ha.

Sulfur Status in the Tropics — Removals

The removal of sulfur depends on sulfur needs of the crops and cropping systems, sulfur losses through drainage and immobilization, efficiency of applied sources of sulfur, and interactions of sulfur with other nutrients.

Crop removal appears to be the major source of depletion of sulfur, and drainage or leaching losses seem to be second. Volatilization loss in submerged soils could be another source of sulfur depletion. However, the nature and magnitude of sulfur depletion depends upon the crop, soil, and other factors. The adsorption of sulfur also reduces the amount of sulfur available to the crop. The average sulfur removed in producing 1 mt of food grain for important crop groups is as follows: cereals (wheat and rice) — 3-4 kg, sorghum and millet — 5-8 kg, pulses and legumes — 8 kg, and oilseeds — 12 kg.

Estimates of average levels of sulfur additions from fertilizer sources and sulfur uptake requirements per hectare of cropped area in India, Indonesia, Nigeria, and Brazil are given in Table 10.3. These estimates indicate that in

Table 10.3. Nutrient consumption, sulfur uptake, and sulfur supply in selected tropical countries during 1970 and 1980.

Year	Country	Nutrient consumption[a] (kg/ha)				Total cropped area[b] ('000 ha)	Sulfur uptake		Sulfur consumption	
		N	P_2O_5	K_2O	Total		'000 mt	kg/ha	'000 mt	kg/ha
1980	Brazil	14.6	32.1	21.1	67.8	43 700	349	8.0	482	11.0
	India	20.8	6.5	3.7	30.9	148 271	784	5.3	250	1.7
	Indonesia	44.4	14.2	4.5	63.0	15 448	130	8.4	48	3.1
	Nigeria	3.0	1.8	0.8	5.7	16 064	66	4.1	17	1.1
1970	Brazil	8.2	11.1	9.0	28.3	32 052	205	6.4	252	7.9
	India	9.0	2.8	1.4	13.2	141 678	651	4.6	346	2.4
	Indonesia	11.1	1.6	0.4	13.1	14 293	87	6.1	c	c
	Nigeria	0.1	0.1	–	0.3	15 849	66	4.2	c	c

a. Per hectare of arable land and permanent crops.
b. Total of the 3-year averages centered on the years shown. The total cropped area includes area under root crops, pulses, oilseeds, wheat, rice, maize, millet, sorghum, sugarcane, soybeans, groundnuts, oil palm, cotton, coffee, and tobacco.
c. Not available.

198

Table 10.4. Sulfur additions, removals, balance and replacement requirements under subsistence and modern agriculture: Likely alternative scenarios. (kg/ha of S)

Sources	Subsistence agriculture		Modern agriculture			Assumptions
	1	2	1	3	3	
Additions						
1. Atmospheric additions (rain-dust-gaseous)	3.0	3.0	3.0	3.0	3.0	kg/ha/annum (means for Nigeria and Kenya are 2.35 and 5.21 kg/ha)
2. Irrigation water						
Rainfed crop	–	–	–	–	–	
Irrigated, 30 cm water	–	2.0	–	–	–	Water[a] containing 2 ppm SO_4-S
Irrigated, 90 cm water	–	–	6.0	6.0	–	Water containing 2 ppm SO_4-S
Irrigated, 90 cm water	–	–	–	–	30.0	Water containing 10 ppm SO_4-S
3. Fertilizers ($N + P_2O_5 + K_2O$)						
15 kg nutrients	1.0	1.0	–	–	–	India's mean in 1970 = 13.2 kg/ha
120 kg nutrients	–	–	8.0	–	–	Approximately equal to mean of Punjab (India) and 2 times that of Brazil and Indonesia
240 kg nutrients	–	–	–	16.0	16.0	Approximately equal to mean of Ludhiana district in Punjab, India
4. Pesticides and chemicals	–	–	–	–	–	Negligible
5. Farmyard manure (FYM)						
1 mt/3 years	0.6	0.6	1.2	–	–	0.2% S in FYM
2 mt/3 years	–	–	–	1.2	1.2	0.2% S in FYM
6. Crop residues	–	–	–	–	–	All removed or burned[b]
Total additions	4.6	6.6	18.2	26.2	50.2	

Table 10.4. Continued.

Sources	Subsistence agriculture		Modern agriculture			Assumptions
	1	2	1	3	3	
Removals						
1. Crops						
L_{00}	4.6	–	–	–	–	L_{00}→yield less than 1 mt/ha (mean yield of India 1970)
L_{01}	–	6.9	–	–	–	L_{01}→irrigated subsistence with average yield 50% higher than in L_{00}
L_1, intensive cropping	–	–	36.0	–	–	L_1→6 mt/ha food grain/year from 2–3 crops
L_2, intensive cropping	–	–	–	72.0	72.0	L_2→12 mt/ha food grain/year from 2–3 crops
2. Drainage or leaching loss	0.6	0.6	1.8	3.6	9.6	½ of estimate of Nigeria and Kenya / Higher leaching because of higher SO_4 content and higher irrigation
3. Adsorbed or immobilized S in irrigation water	0.0	0.5	1.5	1.5	7.5	¼ of S from irrigation water
Total removals	5.2	8.0	39.3	77.1	89.1	
Balance						
Balance (deficit) I	–0.6	–1.4	–21.1	–50.9	–38.9	Similar share of S in fertilizer as in 1980–81 in India (1/15 of nutrients)
Balance (deficit) II	–1.6	–2.4	–29.1	–66.9	–54.9	Completely S-free fertilizers used
Replacement requirements						
Fertilizer S required, I	1.1	2.5	36.9	88.1	68.1	S deficit I×1.75
Fertilizer S required, II	2.8	4.2	50.9	117.1	96.1	S deficit II×1.75
Fertilizer S required, I	2.2	5.0	73.8	176.2	136.2	S deficit I×3.50
Fertilizer S required, II	5.6	8.4	101.8	234.2	192.2	S deficit II×3.50

a. Assuming all the SO_4 remains within the root zone which is not likely.

b. If burned, some SO_4 may be retained by Ca, K, and Mg in ash. However, empirical estimates are not available.

the transitional stage, as agriculture changed from subsistence to modern farming systems, the sulfur requirements per hectare increased and sulfur additions generally decreased. This is a matter of great concern. Similar trends exist in most of the other developing countries. In Brazil, for example, the aggregate sulfur addition through fertilizers seems to match sulfur uptake by crops; yet when one considers the high sulfate-fixing nature of the soil, the higher ratio of P to S (4.5:1), and higher leaching, one cannot be complacent about the sulfur availability. Furthermore, it is possible that not all soils and crops receive sulfur. In order to account for low use efficiency, replenishment of sulfur through fertilizers should be much higher than sulfur uptake by the crop.

There are no good estimates of sulfate losses in leaching or drainage water. However, the experience in Kenya showsthat the annual loss of sulfur under very high rainfall is 2.21 kg/ha, whereas in Nigeria it is 0.3 kg/ha. Thus, for most situations the sulfur loss under subsistence farming may not be more than 0.6 kg/ha.

Sulfur Balance Sheet and Likely Scenarios

On the basis of average sulfur addition and removal estimates, we have developed a sulfur balance sheet and likely scenarios for subsistence and modern agricultural systems in the tropics. These results are reported in Table 10.4. The balance sheet clearly indicates the serious sulfur deficiency problems that are emerging in the tropics and the need for realistic sulfur supply strategies.

Subsistence Agriculture

Under subsistence agriculture the additions and removals of sulfur may leave a slight deficit (0.6 kg/ha), which can be supplied by the soil provided it is not inherently deficient in sulfur (Scenario 1). Otherwise the crop yield will be seriously reduced by sulfur deficiency. The sulfur deficit could increase to 1.6 kg/ha if the fertilizer used is sulfur free.

Even for a farmer trying to produce 1.5 times more food grain per hectare, under subsistence agriculture the sulfur supply deficit will increase from 0.6 to 1.4 kg/ha (Scenario 2). The sulfur deficit will become 2.4 kg/ha if the fertilizer applied in the system does not contain any sulfur. This puts a great strain on the sulfur reserves in the soil or depresses the crop yields.

Modern Agriculture

Modern agriculture is assumed to be based on high-yielding crop varieties, intensive cropping, and high inputs of fertilizers and irrigation. In spite of all the sulfur added incidentally through fertilizers, manure, and irrigation water as calculated for situations closely matching the actual situation in India (N:S ratio of 15:1), the sulfur deficit is estimated to be very high. In three scenarios under modern agriculture, the sulfur deficit is estimated to be 21.1, 50.9, and 38.9 kg/ha, depending upon the level of production, fertilizer use, and sulfate content of irrigation water. The sulfur deficit for these scenarios will rise to 29.1, 66.9, and 54.9 kg/ha if the applied fertilizers have no sulfur in them.

The amount of fertilizer sulfur required to replenish these amounts is also indicated in the table. If the use efficiency of fertilizer sulfur is only 28.5% (replacement coefficient 3.50), i.e., one-half of the assumed use efficiency of 57.14% (replacement coefficient 1.75), the amount of fertilizer sulfur required will be twice as much. It may be further observed that with irrigation water containing 10 ppm sulfate sulfur the sulfur deficit is considerably reduced but not completely eliminated. If the small amount of sulfur being incidentally applied were absent, the situation would be very serious. Under these circumstances, the use efficiency and productivity of fertilizers (NPK) and other inputs such as irrigation water would be greatly reduced.

Such intensive cropping and exploitive agriculture depletes the sulfur reserves of the soil and has an adverse effect on crop yields. According to long-term experiments conducted in India, the available sulfur content of the soil declined under intensive cropping and continuous use of sulfur-free fertilizers in 7 years in an alluvial soil. This is an indication of the situation that will develop elsewhere unless sulfur-containing fertilizers are used to supply sulfur. Such long-term experiments should be monitored to study the changes in sulfur supply and requirements and to formulate sound fertilizer management practices for different soils and cropping systems.

Magnitude of Sulfur Deficiency in Tropical Soils

There is general lack of consistent and accurate data about the extent of sulfur deficiency in the tropics. Available estimates indicate that about 52 million ha of high-base soils (11% of total) and 745 million ha of infertile acid soils (71% of total) in Latin America have a sulfur deficiency problem. Campo Cerrado soils of Brazil, the highlands and eastern plains (Llanos) of Colombia, and the highly weathered volcanic soils of the West Indies and Central America are good examples of sulfur-deficient areas. There is evidence that soils of the upland savannas of Africa have low reserves of total and available sulfur and are likely to become more deficient in sulfur under the present land management system. There is greater likelihood of sulfur deficiency in the savannas of Nigeria than in the forest zone soils.

There are about 452 million ha of sandy soil areas in the semiarid tropics, stretching from Latin America to the savannas of Africa and the alluvial soils of Asia, that also are likely to be deficient in sulfur. Evidence from Brazil and Nigeria confirms that newly cleared tropical lands show sulfur deficiency after a few years of cropping. Estimates of the rate of disappearance of sulfur in organic matter vary; they range from 2% in Nigeria to 10% in Latin America.

There is general consensus that sulfur in organic matter disappears faster than does nitrogen in organic matter; thus, the problem of sulfur fertility management is even more difficult. The lack of systematic studies that correlate sulfur deficiency with differences in soil taxonomic groups precludes an accurate assessment and delineation of the sulfur-deficient areas, but broad conclusions can be drawn about the countries where such sulfur-deficient soils exist. A total of 48 tropical countries (10 in Asia, 23 in Africa, and 15 in Latin America) have been identified as having serious sulfur-deficiency problems.

Determining Sulfur Deficiency in the Tropics

National and international fertilizer and agricultural research organizations concerned with sulfur research in the tropics need to consider the following important conclusions and recommendations.

First, the limitations of the soil tests for determining the availability of sulfur to the plants have been pointed out by many; a combination of soil test and plant analysis seems preferable, and the diagnostic techniques need to be standardized.

Second, most researchers have found monocalcium phosphate solution preferable to other extractants for available sulfur in soils; however, the method for estimating sulfur in the extract needs to be refined, and critical values for different crops need to be established.

Third, the usefulness of pot culture and greenhouse tests to assess the need for fertilizer sulfr is limited because the system may overemphasize the nutrient need as a result of conditions under which the plants are grown. Field experimentation is most reliable for diagnosis of the need for sulfur fertilization.

Fourth, for preliminary screening of different sources of sulfur, greenhouse studies could be valuable, provided the researchers consider the actual soil environments and crops for which the proposed sulfur fertilizer is to be used.

Fifth, lysimetric studies under controlled environments are, no doubt, good for understanding the principles involved, but for solving the field problems and for improving sulfur economy of soil it is better to conduct such studies on the soil in situ where the effect of growing a crop can also be studied simultaneously.

Crop Response to Fertilizer Sulfur

Data on responses to fertilizer sulfur from field experiments are rather limited; the information is mostly confined to areas where, year after year, deficiencies have been observed or where, under the impact of modern agriculture, the full potential of inputs is not being realized because of induced sulfur deficiency. Specific examples of sulfur deficiency and response to applied sulfur are discussed in detail in the previous chapters. Despite the inadequacy of the data, however, the following seven conclusions emerge from results based on field experiments in a number of tropical countries.

First, the deficiency of sulfur in the tropics is widespread, though not so spectacular as nitrogen and phosphorus deficiency. Significant responses to application of sulfate sulfur are expected. In some cases significant increases in crop yield have been obtained in greenhouse studies, and they could be considered as indicative of crop response to sulfur and thus the need for sulfur research in the field. The studies also indicate that responses to fertilizers, specifically to nitrogen and phosphorus, will increase if the limiting factor, sulfur, is supplied.

Second, sulfur as a nutrient has an important place in improving quantity and quality of food production, and thus alleviation of hunger and malnutrition. This is especially important in the developing tropical countries in Asia, Africa, and Latin America.

In Asia, sulfur responses were obtained in the 1970s with medium-to-high doses of fertilizers N, NP, or NPK and improved varieties of cereals (rice or wheat). Marked responses to sulfur were observed in oilseeds (groundnuts, soybeans, rape, and mustard), legume forages such as alfalfa and berseem (Trifolium alexandrinum L.), and potatoes. It appears that most of the work was confined to coarse-textured soils and was done after the introduction of highyielding varieties. Whether the problem is localized or extends to a larger area has not been determined; nor has the question of whether the sulfur application is needed for every crop in the rotation or once for each rotation really been studied.

In Africa, most of the research was done before the introduction of high-yielding varieties, i.e., before the 1970s, and was concerned with commercial crops like cotton, groundnuts, and tea. Since the introduction of high-yielding varieties, or the post-independence period, very little research on sulfur fertilization seems to have been done on any crops and least on food crops. In view of the great food deficit in this region, there is a need for intensive and well-coordinated research to assess th need for sulfur fertilizers for food crops in the region.

In Latin America, most of the research reported in literature relates to improved varieties of rice, maize, soybeans, cotton, coffee, beans, and pasture legumes. Marked responses were obtained in the Campo Cerrado soils of Brazil, the highlands and eastern plains of Colombia, and the highly weathered volcanic soils of Central America.

Third, at present the sulfur deficiency under high-yielding varieties may appear to be confined to those soils that were inherently poor. As the intensity of cropping and level of fertilization increase, sulfur deficiency may become a serious limiting factor, especially because of the decline in sulfur input from high-analysis fertilizers that are free from sulfur or have low sulfur contents.

There are clear indications that in Asia, particularly in India, Bangladesh, and Indonesia, intensive cropping combined with the use of high-yielding varieties and heavy applications of sulfur-free fertilizer may be overstraining the sulfur supply reserves of the soil ecosystem and may be limiting the full potential of new technology. Evidence of this is provided by the data on sulfur uptake and incidental sulfur supply through fertilizers in India and Indonesia. In 1980, the amount of sulfur taken up by field crops in India was estimated to be 784,000 mt, whereas the amount added through fertilizers was hardly 250,000 mt. In Indonesia the comparable figures were 130,000 and 48,000 mt.

It is also possible that irrigation water high in sulfate may counteract this sulfur deficiency in many situations; in other cases, sulfur-containing fertilizers may be needed. Crops like oilseeds, pulses, legumes, and forages, which remove relatively larger amounts of sulfur per metric ton of dry matter than do the cereals, may become more responsive to sulfur fertilizers with the introduction of their high-yielding varieties. Thus, in the future the sulfur problem will become much more serious.

Fourth, a comparison of sulfur supply sources indicates that generally gypsum or other sulfate sources have proved to be effective for most of thesoils and crops. The modifying effects of time and the method and dose of application have also been established in many studies. An even less efficient substance can become an effective sulfur source if the cost:benefit relationship is favorable. However, economic evaluation of sulfur supply sources has generally been ignored. Thus, the evaluation of alternative sulfur supply sources should be based on analytical studies done by interdisciplinary teams of scientists dealing with the technological, agronomic, and economic aspects of fertilizer sulfur research.

Fifth, in most of the tropical countries the sources of sulfur are gypsum, pyrite, or sulfur-containing byproducts of agriculture and industry. Technology needs to be developed for the use of these substances as economic-sources of sulfur for plant nutrition. The technologists, agronomists, and economists must determine whether to modify the fertilizers to incorporate sulfur from these sources or to select compound fertilizers and mixtures for specific situations.

Sixth, long-term studies with tea, coffee, and coconuts have shown that sulfur-containing fertilizers, if continuously applied, build up reserves of adsorbed sulfate in soils which, in turn, reduce the amount of sulfur to be added annually. The results of two long-term experiments, one at Samaru (northern Nigeria) and the other at New Delhi (northern India), clearly indicate changes in sulfur supply from soil under extensive and intensive cropping systems in Africa and Asia, respectively. Long-term studies and monitoring of changes in the status of nutrients in the soils through long-term experiments elsewhere may also be desirable.

Seventh and finally, the experience with tropical agriculture also shows that both phosphate and lime accelerate losses of sulfate sulfur in acid soils that are high in exchangeable aluminum. Thus,the cultivation of such lands requires fertilizer management practices or fertilizer products that can reduce such losses without affecting the usefulness of the lime and phosphate applications.

Priority Areas for Fertilizer Sulfur Research

Though the data are scanty, the published information about sulfur deficiency suggests that the priority areas identified in Table 10.5 are of greatest concern and should be examined to assess the extent and intensity of the sulfur problem, to determine economic sources of sulfur supply, and to formulate appropriate sulfur programs. Research and development programs in these areas would probably have the greatest economic benefits.

The areas where severe sulfur deficiency has been observed should be selected for assessing the magnitude of the problem and the degree of crop responses to applied fertilizer sulfur. A coordinated program of simple fertilizer trials on farmers' fields, based on the missing nutrient concept and using treatments such as control, N, NP, NPK, NPKS, NPKS + micronutrients, might be tried. Soil and plant analyses could be used to supplement such studies. Where possible, ^{35}S should be used.

Sulfur Strategies for Meeting the Challenge

Identification of those factors responsible for creating sulfur deficiency in the tropics is prerequisite to designing strategies for solving the sulfur problem. The more important factors are soil, climate, cropping systems, fertilizer, irrigation, cropping intensity, industrialization, environmental programs, and extension of agriculture to marginal lands.

Although soil and climatic influences are difficult to change, an understanding of them is helpful in the management of sulfur fertilizers. Most serious are the leaching and erosion losses due to rain and adsorption of sulfate under certain conditions. Any strategy for improving the sulfur situation in tropical soils should emphasize the management of soil, fertilizer, and crops.

206

Table 10.5. Priority areas for fertilizer sulfur research and policy in Asia, Africa, and Latin America.

Region	Country	Areas within the country	Crops
Asia	India	Coarse-textured sandy soils of alluvial plains of Punjab, Haryana, Uttar Pradesh, Rajasthan, and certain pockets of Gujarat	Groundnuts, rape & mustard, wheat, maize, chickpeas, soybeans, berseem, potatoes
	Bangladesh	Lowland rice area	Rice, wheat, mustard
	Thailand	Plateau of northeast Thailand	Rice, soybeans, pulses, pasture
	Indonesia	Sulawesi, East Java	Rice, pasture
Africa	Nigeria	Northern Nigeria	Maize, sorghum, roots & tubers, cowpeas, groundnuts
	Senegal	Central and southern Senegal	Groundnuts, cotton, millet, maize
	Kenya	(1) Coastal sandy soils (2) Sandy loam soils of Kitale and Songhor Regions (3) Volcanic soils near Kilimanjaro (4) Bottom lands of Machakos area	Maize, cotton, pastures
	Zimbabwe	Sandy soils	Maize, groundnuts, tea
Latin America	Brazil	Highly weathered soils of Brazilian plateau Campo Cerrado soils of Sao Paulo Region	Maize, rice, cotton, pastures, soybeans, coffee
	Colombia	Bogota Highlands Eastern plain (Llanos)	Maize, soybeans, beans, pasture, legumes, coffee

In some areas, because of the high sulfate content, the irrigation water could supply the sulfur needs of the crop. In other areas, where waters of low sulfate content are used, sulfates may be leached and the soil impoverished. Analytical studies dealing with the effects of irrigation, fertilizer application, and cropping system combinations under field conditions are essential to understand the phenomena.

Intensification of agriculture, use of high-yielding crop varieties, and increasing use of fertilizers low in sulfur are aggravating the sulfurdeficiency problem. Corrective measures need to be taken through use of fertilizers with higher sulfur content to lower the N:S ratio or P:S ratio in the fertilizers.

From the long list of sulfur-containing substances it appears doubtful that the use of fluid fertilizers will be feasible in any significant part of the developing tropical countries in the ear future. However, dry fertilizers rich in sulfate may become more attractive. The need for modified, economically efficient fertilizer technology is of high priority. Sulfur-fortified triple superphosphate, concentrated superphosphate, partially acidulated phosphate rock, and sulfate-enriched diammonium phosphate may become more acceptable. Appropriate technologies need to be developed to supply them.

Strategies to supply sulfur to soils and crops in every country should be based on the use of waste products from the fertilizer industry, chemical industry, and agricultural industry; on local sulfur resources; or on incorporation of sulfur in the popular fertilizer products. Ammonium sulfate and single superphosphate have important places in the agriculture of these countries, and strategies can be developed to make their use more efficient and economical. Use of sulfur-coated urea and sulfur-enriched fertilizers on sulfur-deficient soils and for certain crops needs to be given a fair trial.

Gypsum, elemental sulfur, and pyrites will be preferable for alkaline and calcareous soils of the semiarid tropics. However, for highly acid soils the use of elemental sulfur and other acid-forming substances is questionable. More emphasis should be placed on the use of phosphogypsum directly or through incorporation in fertilizers. In fact, there is a need for interdisciplinary research by teams of fertilizer technologists, agronomists, and economists to develop suitable fertilizer products that supply sulfur for different soils and cropping systems while keeping in view the locally available sulfur resources.

Estimating Sulfur Requirements and Gaps

In order to design nationally acceptable and economically viable supply strategies to correct sulfur deficiency and improve the production and quality of food, there is a need to accurately estimate the sulfur requirements, sulfur supplies, and implied sulfur gaps.

Sulfur estimates are based on simple statistical analysis. The sulfur requirement estimates are divided into two broad categories. These are (1) requirements by field crops for sulfur as a nutrient and (2) sulfur requirements in the fertilizer industry. The requirements for sulfur as a plant nutrient for crops are based on (a) crop production levels, which account for the area under different crops and average crop yields; (b) average sulfur uptake by crops; and (c) use efficiency of applied sulfur. In this context the estimated sulfur requirements do reflect the impact of factors such as changing cropping patterns, expanding crop yields, and multiple cropping.

The requirement for sulfur as a plant nutrient is further classified into three categories. These are (1) sulfur uptake; (2) sulfur replacement I requirements that assume a sulfur-replacement coefficient of 1.75 which implies sulfur use efficiency of approximately 60%; and (3) sulfur replacement II requirements that assume a sulfur replacement coefficient of 3.50 which implies sulfur use efficiency of one-half, approximately 30%.

The sulfur uptake and the replacement requirements are based on actual crop production from 1960 to 1980 and on projected crop production from 1985 to 2000. The results refer to 3-year simple averages centered on years shown in order to avoid the effect of weather-related variations in crop production. On the other hand, sulfur requirements in the fertilizer industry are based on fertilizer production needed to meet projected fertilizer consumption requirements from 1980/81 to 2000/01. Estimated sulfur requirements in the fertilizer industry are based on sulfur needed to manufacture ammonium sulfate, single superphosphate, triple superphosphate, and ammonium phosphates.

The estimates of sulfur requirements are based on production of 25 important field crops. These crops are then divided into 15 specific crop groups including wheat, rice, maize, millet, sorghum, pulses, oilseeds, soybeans, grondnuts, oil palm, root crops, sugarcane, cotton, coffee, and tobacco. Each country under study does not necessarily grow all these crops. Furthermore, the relative importance of individual crops varies from one country to another. These 25 crops, however, account for most of the chemical fertilizer consumption in developing countries of the tropics.

The estimates for sulfur requirements are made for 11 tropical countries and 3 regions in addition to the world as a whole. The 11 countries include India, Indonesia, and the Philippines from Asia; Kenya, Niger, Nigeria, Sudan, and Zimbabwe from Africa; and Brazil, Colombia, and Mexico from Latin America. The three regions include the Far East, Africa, and Latin America (as developing market economies). The crop production levels are specific to each country and region.

However, average sulfur uptake by crop groups broadly represents developing tropical countries and is assumed to be the same for all the countries and regions. Finally, the assumed two levels of sulfur use efficiency (replacement coefficients) are not specific to any crop, country, or fertilizer; rather, they represent typical conditions prevailing in temperate and tropical countries. In this context, the estimated sulfur requirements presented here are good first-order approximations.

The estimated aggregate sulfur requirements by field crops and the world fertilizer industry are summarized in Tables 10.6 and 10.7. India and Brazil

209

Table 10.6. Estimated aggregate sulfur requirements for field crops in selected developing
tropical countries and regions, 1960 – 2000.

Country/region	Sulfur requirements[a] ('000 mt of S)								
	Uptake			Replacement I			Replacement II		
	1960	1980	2000	1960	1980	2000	1960	1980	2000
Asia									
India	509	784	1 333	891	1 372	2 332	1 782	2 743	4 665
Indonesia	64	130	294	112	227	514	225	454	1 028
Philippines	21	45	108	38	80	190	75	159	379
Africa									
Kenya	9	15	21	15	26	37	31	52	73
Niger	6	13	29	11	22	51	22	44	103
Nigeria	56	66	98	99	116	171	197	231	343
Sudan	29	30	64	51	53	113	102	107	225
Zimbabwe	7	16	29	12	29	51	25	58	101
Latin America									
Brazil	157	349	708	275	611	1 239	550	1 222	2 477
Colombia	21	38	86	36	67	150	72	134	301
Mexico	64	137	240	113	239	419	225	478	838
Africa	224	327	411	392	573	720	783	1 146	1 440
Far East	776	1 340	2 463	1 358	2 345	4 310	2 717	4 690	8 620
Latin America	404	844	1 603	707	1 477	2 805	1 413	2 954	5 610
World	4 582	7 911	14 238	8 018	13 844	24 917	16 037	27 687	49 835

a. Uptake: Sulfur uptake by field crops; replacement I: uptake * sulfur replacement coefficient of 1.75 (implies use efficiency of about 60%); and replacement II: uptake * sulfur
replacement coefficient of 3.50 (implies use efficiency of about 30%).

Table 10.7. Estimated world sulfur requirements as crop nutrient and for the fertilizer industry, 1960 – 2000.

Type of sulfur requirements	Sulfur requirements during (million mt of S)		
	1980	1990	2000
Sulfur as crop nutrient			
Uptake	7.9	10.5	14.2
Replacement I	13.8	18.4	24.9
Replacement II	27.7	36.9	49.8
Sulfur for fertilizer industry			
AS/SSP	6.5	10.0	16.3
TSP/AP[a]	20.0	31.1	48.4
Total	26.5	41.1	64.7

a. AP = ammonium phosphate.

stand out since the largest sulfur requirements are in Asia and Latin America, respectively. With the exception of Mexico, all the other countries studied are net sulfur importers. The sulfur requirements in the world fertilizer industry are much more than the sulfur uptake. However, only a small percentage of sulfur used in the fertilizer industry is transferred to the agricultural sector through sulfur-containing fertilizers.

The estimated proportion of sulfur uptake by cereal and noncereal food crops is shown in Figure 10.1. A large share of sulfur uptake is attributed to cereal crops. However, the share varies not only by country or region but also over time. A shift in the cropping pattern in favor of those crops that, on the average, remove relatively large quantities of sulfur would increase aggregate sulfur uptake and hence the fertilizer sulfur replacement requirements. The production of soybean in Brazil highlights such a transformation.

The estimated aggregate sulfur requirements, supply, and gaps for selected countries during 1980 are reported in Table 10.8 and for India from 1960 to 2000 in Figure 10.2. The results indicate that, except for Mexico, the sulfur gaps are rather large in relation to the current fertilizer sulfur supply. Furthermore, the projected sulfur gaps are estimated to increase from 1980 to 2000. For example, it is estimated that sulfur gap III (fertilizer sulfur supply minus sulfur replacement II requirements) in India will increase from 2.5 million mt in 1980 to 3.1 million mt in 1990, and to 4.0 million mt in the year 2000.

The models used in this study provide a systematic means for assessing the sulfur requirements for major agricultural crops, sulfur supply and the potential sulfur gaps that may prevail. While the results indicate that large sulfur gaps now exist or soon will emerge in a number of developing countries, the authors wish to point out that these estimates are first-order ap-

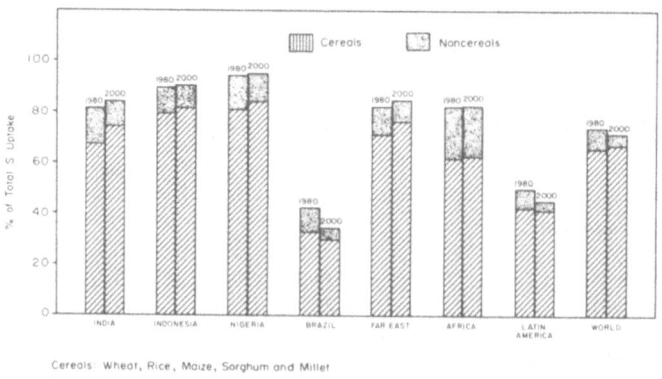

Figure 10.1. Estimated Proportion of Sulfur Uptake by Cereal and Noncereal Food Crops in Selected Countries and Regions in the World.

Table 10.8. Estimated sulfur requirements, fertilizer sulfur supply, and sulfur gaps in select-ed tropical countries during 1980. ('000 mt of S)

Country	Sulfur requirements			Fertilizer sulfur supply	Sulfur gaps[a]		
	Uptake	Replace-ment I	Replace-ment II		I	II	III
Asia							
India	784	1 372	2 743	250	− 534	− 1 122	− 2 493
Indonesia	130	227	454	48	− 82	− 179	− 406
Philippines	45	80	159	30	− 15	− 50	− 129
Africa							
Kenya	15	26	52	11	− 4	− 15	− 41
Niger	13	22	44	<1	− 12	− 21	− 43
Nigeria	66	116	231	17	− 49	− 99	− 214
Sudan	30	53	107	b	b	b	b
Zimbabwe	16	29	58	b	b	b	b
Latin America							
Brazil	349	611	1 222	482	+ 133	− 129	− 740
Colombia	38	67	134	5	− 33	− 62	− 129
Mexico	137	239	478	338	+ 201	+ 99	− 140

a. Fertilizer sulfur supply minus sulfur requirements. Gap I is supply minus uptake; Gap II
   is supply minus Replacement I; and Gap III is supply minus Replacement II.
b. Not available.

proximations. Although the sulfur gap estimates are based on the best avail-able data, much work remains to be done for further refinements in the context of specific countries. Recognizing these limitations, the results have major implications for national fertilizer policy.

These large sulfur gaps have important policy implications for fertilizer research, sulfur supply, fertilizer material selection, fertilizer imports, fer-tilizer distribution, sulfur promotion, capital investment, and foreign ex-change allocation. Unless something is done to bridge these gaps and to correct the sulfur-deficiency problem, national and international efforts to accelerate food production will be seriously handicapped. This is a real challenge and a great opportunity for all those involved in fertilizer policyformulation, fertilizer research, and fertilizer production, distribu-tion, and use to make a major contribution to food production of develop-ing tropical countries.

Evaluation of Alternative Fertilizer Sulfur Materials

In order to deal with increasing sulfur deficiencies and projected sulfur

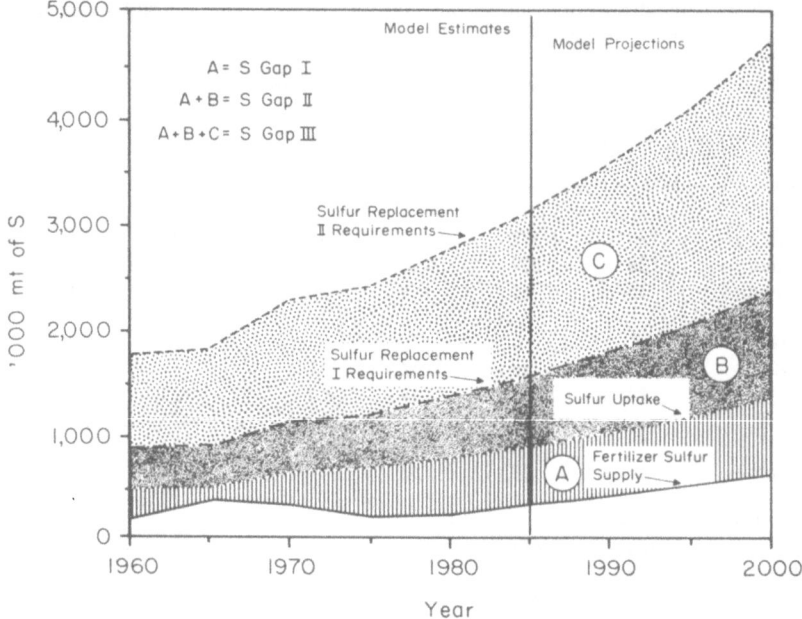

Figure 10.2. Estimated Sulfur Requirements, Fertilizer Sulfur Supply, and Sulfur Gaps in India: An Example.

gaps, there is a need to identify, develop, evaluate, and transfer fertilizer sulfur technology and strategies that would be appropriate, technically and economically, for tropical countries. Alternative sulfur supply strategies include the use of all or some of (1) conventional sulfur-containing fertilizers; (2) modified sulfur-containing fertilizers; and (3) indigenous sulfur supply sources such as gypsum, phosphogypsum, and pyrites.

In developing and recommending any of these sulfur-containing fertilizers, it is extremely important to keep in mind the climatic conditions, cropping systems, soil types, socioeconomic conditions, and natural resource endowments of those tropical countries that are the target of such technology. All the existing and modified sulfur-containing fertilizer technologies must be evaluated with respect to the following criteria:

1. The technical feasibility of production, distribution, and use of sulfur-containing fertilizers.
2. Agronomic effectiveness under farmers' field conditions.
3. Preferences and general attitudes of those involved in production, distribution, and use of these materials.
4. Economic effectiveness under free-market conditions.
5. Economic effectiveness under prevailing government policy with respect to fertilizer sulfur.

6. Foreign exchange use, earnings, and savings.
7. Economic and financial aspects of research, production, distribution, and use of these materials.
8. Existing and suggested government policies dealing with fertilizer sulfur raw materials, production, distribution, promotion, regulation, pricing, subsidies, trade, and research.

The technology for each proposed sulfur-containing fertilizer needs to be evaluated within an interdisciplinary context from the moment a technology is conceived until it is ready for transfer and general use by farmers in the tropical countries.

Economic and Policy Analysis of Fertilizer Sulfur

Despite the need for sulfur as an essential plant nutrient and the sustantial returns expected from its use, very little analytical or empirical research has been done on the economic and policy analyses of fertilizer sulfur in tropical countries. There are several reasons for this situation. The primary reason, however, is that the economic importance of fertilizer sulfur has not been recognized. In order to formulate appropriate sulfur policy at the national level, there is a need for the following:

1. Determination of economic returns to sulfur use under different agroclimatic conditions and cropping systems.
2. Comparative economic evaluation of existing, modified, and indigenous sulfur-containing fertilizers and amendments.
3. Economic evaluation of phosphogypsum as a source of sulfur for plant nutrients and raw material for manufacturing sulfuric acid and nitrogen, phosphorus, and potassium fertilizers.
4. Economic analysis that accounts for the residual effects of sulfur, the interaction of sulfur with other nutrients, and the productivity of fertilizer sulfur under different crop technologies and cropping systems.
5. Determination of the delivered price of sulfur to farmers.
6. Economic evaluation of price and the transportation subsidy that must be paid by the government on sulfur-containing fertilizers and other sulfur supply sources.
7. Economic evaluation of sulfur-containing fertilizers from indigenous sulfur supply sources as opposed to imported sulfur or sulfur-containing fertilizers.

Pricing of Fertilizer Sulfur

Sulfur is not a free commodity. The international sulfur prices, expressed in constant dollar terms, generally declined from 1955 to 1978, with the exception of two price upswings that peaked during 1968 and 1975. In any case, the current sulfur prices during this period were less than $50/mt. However, the market changed in 1978 when international dry sulfur prices (f.o.b. U.S. Gulf) started rising from approximately $60/mt in January 1979 to $135/mt in January 1981 and then declined to $112/mt in January 1983 and to $88/mt in October 1983 and then started rising again to $103/mt in January 1984 and to $146/mt in December 1984. Clearly, sulfur prices play an important role in determining the appropriate sulfur supply strategy and prices for sulfur-containing fertilizers.

Sulfur-containing chemical fertilizers, including popular fertilizers such as ammonium sulfate and single superphosphate, are rarely priced for their sulfur content. As long as sulfur in chemical fertilizers is considered merely a bonus, the fertilizer industry will have little economic incentive to manufacture sulfur-containing fertilizers as main products.

In those areas facing serious sulfur deficiency, the returns to the use of sulfur-containing fertilizers priced for sulfur appear extremely favorable, both at the farm and national levels. However, sulfur pricing does raise extremely important policy questions that can be resolved only after sound economic and policy analyses. Some of these questions are as follows:

1. How should the production costs be allocated and the fertilizer be priced when the sulfur-containing fertilizer is produced as a coproduct or as a byproduct?
2. How should the price of sulfur-containing fertilizers be determined at the retail level in comparison with other competitive sulfur-free fertilizers?
3. Would the farmers be willing to pay for sulfur in sulfur-containing fertilizers?
4. How would the fertilizer industry react to the creation of new capacity to produce sulfur-containing fertilizers?
5. How would the costs and benefits of the sulfur-pricing policy be distributed among producers, distributors, farmers, and consumers?

Sulfur Situation, Resources, Trade, and Outlook

Sulfur is considered one of the more abundant elements on earth. However, only a small fraction of large sulfur resources is recoverable at current levels of knowledge, technology, and prices. The key points concerning the sulfur industry as it relates to the fertilizer sector and tropical agriculture are analyzed below:

First, during 1981 world production of sulfur was about 53.8 million mt. Four countries (United States 23.7%, U.S.S.R. 18.0%, Canada 12.6%, and Poland 9.2%) accounted for almost two-thirds of world sulfur production. Approximately 63% of the estimated production was brimstone, and this proportion is expected to increase because of increased production of recovered sulfur. During 1981 the estimated share of developing tropical regions in world sulfur production was only 7.8%, and Mexico alone accounted for 50% of that production (Table 10.9). However, the oil-producing countries, especially in the Middle East, are expected to increase their production of recovered sulfur.

Second, a large share of sulfur (80%-90%) is used to manufacture sulfuric acid. The share of developing tropical regions in worldwide sulfur consumption (55.0 million mt) was only about 12.4% during 1980. This is rather a small amount in the context of their relative share in world population, agricultural production, and food needs. During 1980, the share of individual developing tropical regions in world sulfur consumption was estimated to be 3.3% in Africa, 4.6% in Asia, and 4.5% in Latin America.

Third, during 1981 world production of sulfuric acid was 138 million mt. Four countries (United States 26%, U.S.S.R. 17%, China 6%, and Japan 5%) accounted for 54% of world sulfuric acid production. Among the developing tropical countries, the major producers of sulfuric acid in 1981 were India 2%, Mexico 2%, Brazil 1.7%, Morocco 1.7%, and Tunisia 1.6%. The developing market economies as a group accounted for about 13%.

Fourth, for the world as a whole, 55% of the sulfuric acid was used in the fertilizer industry during 1981. The share allocated to the fertilizer industry varies from 45% in Asia (including Japan) to 67% in Latin America and 75% in Africa (including South Africa). A large share (90% in the western world) of the sulfuric acid used in the fertilizer industry goes to manufacture phosphate fertilizers. Most of the sulfur used in the phosphate fertilizer industry (with the exception of single superphosphate) is discarded in byproduct phosphogypsum.

Table 10.9 Sources of sulfur and sulfur production in developing tropical regions of the world during 1981.

Region	Production ('000 mt)	Share in world production (%)	Sources of sulfur, % share			
			Brimstone	Pyrites	Other forms	Total
Africa[a]	178	0.3	6	27	67	100
Asia[b]	1 335	2.5	74	8	18	100
Latin America	2 699	5.0	90	3	7	100
Regions total	4 212	7.8	82	5	13	100
World	53 727	100.0	63	21	16	100

a. Excluding South Africa with a total production of 636 thousand mt.
b. Excluding Japan with a total production of 2 706 thousand mt. China is also excluded.

Fifth, during 1981 about 30% (16 million mt) of world sulfur production was traded in the international market, most of it in the form of brimstone. Canada and Poland accounted for about 70% of brimstone exports. Among the developing countries, Mexico is the only net exporter of sulfur. Potential sulfur exporters include countries from the Middle East. With the exception of Mexico, most of the developing countries are net importers of sulfur, which is esential to develop their agricultural and industrial sectors.

Sixth, the amount of known sulfur reserves in the world is estimated to be between 1.8 and 2.2 billion mt, which does not include vast sulfur resources contained in anhydrite, gypsum, and sea water. Of these reserves the developing tropical countries are estimated to account for about 33%, including 5% in Mexico, 8% in Iraq, 14% in other countries in the Near East, and 6% in other developing tropical countries of Asia, Africa, and Latin America.

The developing tropical countries are also known to have sulfur resources in the form of anhydrites, gypsum, pyrites, and nonferrous sulfides, as well as elemental sulfur deposits. The following important questions need to be addressed: (1) What is the agronomic effectiveness and technoeconomic feasibility of these sulfur resources? and (2) How should the indigenous sulfur resources be used to meet the sulfur requirements of developing tropical countries in Asia, Africa, and Latin America?

Sulfur Trade Policies

Most tropical countries with serious sulfur-deficiency problems are also net importers of sulfur, usually brimstone. The available information indicates that most of these countries also impose some form of trade restrictions, especially tariffs, on imported sulfur.

Tariffs may be good sources of government revenue, but they can also result in high social costs. The immediate impact of a tariff is that the price of imported sulfur goes up and the increase is reflected in the price of fertilizers such as phosphate fertilizers that use sulfur or sulfuric acid in their manufacture. Tariffs are generally imposed under the pretext of protecting the domestic industry (infant industry argument), but they also encourage inefficiency. For a low-income country with widespread sulfur deficiency, tariffs on sulfur imports may not be in the best interests of the general population since they encourage both inefficiency and inequity.

Phosphogypsum: A Source of Fertilizer Sulfur

Phosphogypsum, a byproduct of wet-process phosphoric acid, is generally considered a waste product, even though it contains all of the sulfur from the sulfuric acid used in the process. The cumulative world production from 1967 to 1985 is estimated to be 1.6 billion mt of phosphogypsum, which includes 265 million mt of sulfur; of this amount, 37 million mt of sulfur equivalent will have been produced in the tropical countries of Asia, Africa, and Latin America.

While large stocks of phosphogypsum are accumulating, the crops in many tropical countries are suffering from sulfur deficiency. Naturally, these countries can ill afford to throw away phosphogypsum, and hence imported sulfur, especially when it was bought with scarce foreign exchange. Furthermore, the technical and economic problems associated with correcting sulfur deficiencies may not be serious at all when phosphogypsum is used as a source of fertilizer sulfur in agriculture. The use of phosphogypsum in agriculture may be a potential source of both sulfur and calcium for increased food production.

There is a need to initiate research to determine (1) the agronomic effectiveness of phosphogypsum as a source of sulfur; (2) the technical problems related to drying, transporting, storing, handling, and converting phosphogypsum to ammonium sulfate and sulfuric acid and for upgrading popular fertilizers with sulfur or sulfur compounds; and (3) the economic viability of phosphogypsum as a raw material source for manufacturing modified nitrogen, phosphorus, and potassium fertilizers, mixtures, and compounds.

Need for Information Related to Fertilizer Sulfur

Appropriate information about sulfur supply, use, response, uptake, prices, and economics is extremely vital in the formulation of policies concerning fertilizer sulfur. Yet most national and international organizations engaged in collecting and publishing information do not include sulfur in fertilizer-related statistics. Consequently, most of this information is not available to either policymakers or researchers.

Sulfur data must be made an integral part of national fertilizer data collection systems. These data must include information related to sulfur raw materials, reserves, and resources; sulfur production, consumption, and trade; sulfur freight and prices; sulfur uptake by crops; and sulfur response. International organizations such as FAO, IFDC, and UNIDO can play an important role in stimulating such programs at the national level, particularly in developing countries.

Fertilizer Sulfur Use Recommendations

With the exception of a few isolated examples, fertilizer recommendations do not include sulfur as one of the plant nutrients. Only the primary nutrients are generally included in the fertilizer recommendations. Any effort directed at correcting sulfur deficiency must involve a set of recommendations that include the use of sulfur at the farm level.

It is extremely important to develop sulfur recommendations based on farm-level data under actual farming conditions. These recommendations must be crop-specific and must also specify the amount to use, the time of application, the method of application, and the source of fertilizer sulfur. Furthermore, the recommendations must be based on sound information regarding crop response to applied fertilizer sulfur and economics of sulfur use.

Fertilizer Sulfur Regulation and Labeling

Lack of use of fertilizer sulfur in countries with severe sulfurdeficiency problems results in a high social cost in terms of lost agricultural production. Under these circumstances, it is economically justifiable to implement government regulations with respect to sulfur supply. These regulations should ensure a fertilizer sulfur supply at the retail level through the availability of popular sulfur-containing fertilizers and/or the requirement that the different fertilizer materials contain a certain minimum amount of sulfur.

The existing quality control regulations in most countries require labeling of only the primary plant nutrients, N, P_2O_5, and K_2O, on fertilizer bags. This is true even when the fertilizer also contains sulfur. The labeling of fertilizer bags with sulfur contents would provide additional information to the farmer who purchases the fertilizer. The additional cost for sulfur labeling is expected to be rather negligible.

Sulfur, Environmental Protection, and Food Production

In the industrialized countries environmental pollution, partly caused by sulfur dioxide and acid rain, is at the center of much public debate. These countries have various laws restricting the emissions of sulfur dioxide to the atmosphere. However, in most of the nonindustrialized developing countries environmental pollution, primarily by sulfur dioxide emission, is not a serious problem. Even in industrialized countries sulfur dioxide is not solely responsible for acid rain although problems may be more serious around localities with industrial complexes.

Obviously there is a need for soe restrictions on emissions of sulfur dioxide and other industrial pollutants in order to protect the environment and public health. However, sulfur dioxide emission standards that are too strict may be counterproductive. Atmospheric sulfur is an important source of sulfur as a plant nutrient. Further restriction on sulfur dioxide would reduce the sulfur supply from the atmosphere at a time when sulfur replacement requirements as a plant nutrient alone are expected to increase worldwide from 28 million mt in 1980 to about 50 million mt in the year 2000.

Furthermore, extremely restrictive sulfur dioxide emission standards would add to social costs in four different ways. These added social costs include (1) cost of enforcing regulations, (2) capital investment in equipment to reduce sulfur dioxide emissions, (3) cost of fertilizer sulfur to supply sulfur that would otherwise be missing as a plant nutrient, and (4) loss in agricultural production if the loss from atmosphere is not made up through alternative sulfur supply sources. Clearly, there is a tradeoff between environmental protection and food production, and environmental policy must be based on a careful analysis of costs, benefits, and the distribution of costs and benefits.

Implications for Research and Public Policy

It must be recognized that sulfur deficiency is either inherent or being induced. There is a widening gap between the sulfur supplied to the soil and that withdrawn from it as a result of a changing agricultural system that involves the use of high-yielding crop varieties, intensive cropping, and the increasing use of sulfur-free fertilizers. The problem calls for a high priority on research and development programs by the national and international organizations.

Priorities for National Research Programs

The national research institutions and policymakers should immediately proceed as follows:
1. Recognize that sulfur deficiency may be limiting crop production and adversely affecting the quality of agricultural production, as well as the health of animals and human beings. The sulfur problem is likely to become more serious in the future; thus, it calls for an immediate, appropriate coordinated action.
2. Identify soils that are deficient in sulfur, using soil and plant tissue testing methods, and give high priority to coarse-textured soils, intensively cropped soils, highly weathered soils, and old volcanic ash soils, i.e., Ultisols, Oxisols, Alfisols, Andepts, and the Inceptisols.
3. Organize coordinated simple fertilizer trials on farmers' fields to study responses to sulfur. The sulfur-responsive crops and their improved varieties should be used on soils identified as deficient or thought to be deficient or responsive to sulfur, with or without lime. It should be

recognized that correction of the acidity, the use of nitrogen, phosphorus, and potassium fertilizers, and other factors affecting nutrient availability are essential to getting the best results from sulfur application.

4. Encourage research on the dynamics of sulfur applied to soils through fertilizers and manures in long-term experiments for agriculture based on high-yielding crop varieties and intensive cropping.

5. Assess the sulfate content of irrigation water and its contribution to the sulfur status of soil, crops and nutrition, and sulfur losses in drainage waters.

6. Monitor sulfur accretion to the ecosystem from the atmosphere through rain, dust, and gaseous deposition at a few selected sites representative of the major agricultural areas.

7. Identify local sources of sulfate sulfur, characterize their chemical attributes, determine their supply status, and develop a strategy for their use as economic sources of sulfur-containing fertilizers.

8. Develop a strategy for the use of such byproducts of the fertilizer industry as phosphogypsum to enrich the nitrogen, phosphorus, and potassium fertilizers with sulfur.

9. Develop strategies and economic policies for encouraging the production, distribution, and use of sulfur-containing fertilizers or soil amendments in order to improve crop yields and quality.

10. Organize workshops and seminars to collect, assess, and disseminate information on the problem of fertilizer sulfur and its implications for increasing food production and improving human nutrition.

Priorities for International Research Programs

International research organizations should establish the following priorities:

1. Recognize that sulfur problems exist and can be solved through timely action by research and development agencies concerned with food production and nutrition.

2. Improve fertilizer technology to reduce costs for the production of high analysis fertilizers that incorporate 5%-10% sulfur.

3. Develop technology for improving use efficiency of sulfur applied to crops in tropical environments, using sulfur-deficient soils and sulfur-responsive crops and crop varieties.

4. Standardize the chemical methodology for analyzing sulfur in soil and plants and standardize the technique used to study sulfur problems of tropical countries.

5. Realize the serious limitations of greenhouse and laboratory studies of sulfur, and therefore place more emphasis on field studies. The greenhouse studies should investigate principles and determine the relative use efficiencies of various test materials in order to form the basis for

field experimentation, but they should not be considered the end point of research.

6. Establish an international network of field trials to study the effect of sulfur-containing fertilizers on the yield and quality of tropical crops, using high-priority areas and selected crops, varieties, and cropping systems. In this research ^{35}S may be used at a few selected sites where facilities for such work provide the necessary support.

7. Determine the economic viability and the farmers-021- preferences for various fertilizer products designed to supply sulfur to major agricultural areas in the tropics.

8. Formulate fertilizer sulfur-related economic policies appropriate for the developing tropical countries in order to accelerate food production through judicious production, trade, distribution, and use of fertilizer sulfur.

9. Arrange international workshops, seminars, and symposiums for exchange of information and for planning a coordinated program of sulfur research with the scientists in the developing countries of the tropics, and provide training facilities as needed.

10. Assist developing countries in finding alternative technological options to be included in strategies for improving sulfur nutrition of crops, increasing food production, and alleviating hunger and malnutrition.

References

Acharya, N. 1973. 'A Radio-Tracer Investigation of the Effect of Sulphate and Phosphate Application on Paddy,' Indian Journal of Agricultural Chemistry, 6(1):23-32.

Acharya, N., and B. V. Subbiah. 1971. 'Radio-Tracer Investigation on the Effect of Application of Sulphate and Phosphate on the Yield and Uptake of Sulphur and Phosphorus by Cotton,' Indian Journal of Agricultural Chemistry, 4:65-72.'

Agarwal, M. R. 1982. 'Techno Economics of Phospho Gypsum Utilization,' Fertiliser News, 27(12):83-89.

Aiyar, S. P. 1945. 'A Chlorosis of Paddy (Oryza sativa L.) due to Sulphate Deficiency,' Current Science (India), 14:10-11.

Alam, S. M., and M. Karim. 1972. 'Effect of Sulphur on S-Uptake and Dry Matter Yield of Rice Plant with Respect to Soil Sulphur,' Pakistan Journal of Scientific Research, 24(3/4):222.

Amaya, U. M. 1981. "Importancia del Azufre en la Producción Agrícola," El Uso del Azufre para el Desarrollo y Modernización de la Agricultura en América Latina, 1.er Simposio, México, D.F., México.

Anderson, A. J., and D. Spencer. 1950. 'Molybdenum in Nitrogen Metabolism of Legumes and Nonlegumes,' Australian Journal of Scientific Research, 3B:414-430.

Andrew, C. S. 1975. 'Evaluation of Plant and Soil Sulphur Tests in Australia,' IN Sulphur in Australasian Agriculture, K. D. McLachlan, (Ed.), pp. 196-200, Sydney University Press, Sydney, Australia.

Andrew, C. S., B. J. Crack, and C. E. Rayment. 1974. 'Queensland,' IN Handbook on Sulphur in Australian Agriculture, K. D. McLachlan (Ed.), pp. 55-68, Commonwealth Scientific and Industrial Research Organization, Melbourne, Australia.

Andrews, A. C., and D. Manajuti. 1980. 'Effects of Plant Density and P and SFertilizers on the Yield of Pigeon Peas, Cajanus cajan, at Pa Kia,' Thailand Journal of Agricultural Science, 13(1):9-14.

Ansari, A. Q., and D.J.F. Bowling. 1972. 'Measurement of Trans-root Electrical Potential of Plants Grown in Soil,' New Phytologist, 71(1):111-117.

Arora, C. L., V. K. Nayyar, N. T. Singh, and B. Singh. 1983. 'Sulphur Deficiency in Wheat Crop Grown in Soils Under Groundnut-Wheat Rotation,' Fertiliser News, 28(2):27-28.

Arora, S. K., and Y. P. Luthra. 1971. 'Relationship Between Sulphur Content of Leaf With Methionine, Cystine, and Cysteine Contents in the Seeds of Phaseolus aureus L. as Affected by S, P, and N Application,' Plant and Soil, 34(1):91-96.

Aulakh, M. S., and G. Dev. 1976. 'Profile Distribution of Sulphur in Some Soil Series of Sangrur District, Punjab,' Journal of the Indian Society of Soil Science, 24(3):308-313.

Aulakh, M. S., and G. Dev. 1977. 'Comparison of Different Plant Analyses Indexes in Evaluating Sulphur Nutrition of Alfalfa,' Indian Journal of Agricultural Chemistry, 10(1/2):163-171.

Aulakh, M. S., and G. Dev. 1978. 'Interaction Effect of Calcium and Sulphur on the Growth and Nutrient omposition of Alfalfa (Medicago sativa L. Pers.) Using ^{35}S,' Plant and Soil, 50(1):125-134.

Aulakh, M. S., and N. S. Pasricha. 1977. 'Interaction Effect of Sulphur and Phosphorus on Growth and Nutrient Content of Moong (Phaseolus aureus L.),' Plant and Soil, 47(2):341-350.

Aulakh, M. S., and N. S. Pasricha. 1978. 'Interrelationships Between Sulphur, Magnesium, and Potassium in Rapeseed. II. Uptake of Mg and K and Their Concentration Ratio,' Indian Journal of Agricultural Sciences, 48(3):143-148.

Aulakh, M. S., and N. S. Pasricha. 1979. 'Response of Gram (Cicer arietinum L.) and Lentil (Lens cultinaris) to Phosphorus as Influenced by Applied Sulphur and Its Residual Effect on Moong (Phaseolus aureus L.),' Bulletin, Indian Society of Soil Science, 12:433-438.

224

Aulakh, M. S., G. Dev, and B. R. Arora. 1976. 'Effect of Sulphur Fertilization on the Nitrogen-Sulphur Relation in Alfalfa (Medicago sativa L. Pers.),' Plant and Soil, 45(1)75-80.

Aulakh, M. S., N. S. Pasricha, and G. Dev. 1977. 'Response of Different Crops to Sulphur Fertilization in Punjab,' Fertiliser News, 22(9):32-36.

Aulakh, M. S., N. S. Pasricha, and N. S. Sahota. 1977. 'Nitrogen-Sulphur Relationships in Brown-Sarson and Indian Mustard,' Indian Journal of Agricultural Sciences, 47(5):249-253.

Aulakh, M. S., N. S. Pasricha, and N. S. Sahota. 1980a. 'Comparative Response of Groundnut (Arachis hypogaea L.) to Three Phosphatic Fertilizers,' Journal of the Indian Society of Soil Science, 28(3):342-346.

Aulakh, M. S., N. S. Pasricha, and N. S. Sahota. 1980b. 'Yield Nutrient Concentration and Quality of Mustard Crops as Influenced by Nitrogen and Sulphur Fertilisers,' Journal of Agricultural Science, Cambridge, 94:545-549.

Aulakh, M. S., B. Singh, and B. R. Arora. 1977. 'Effect of Sulphur Fertilisation on the Yield and Quality of Potatoes (Solanum tuberosum L.),' Journal of the Indian Society of Soil Science, 25(2):182-185.

Ayala, H. F., R. R. Guerrero, and J. J. Gamboa. 1973. 'Estudio del Azufre en Suelos de Narino y Putu Mayo (Colombia),' Anales de Edafologia y Agrobiologia, 32(3-4):401-416.

Aylmore, L.A.G., M. Karim, and J. P. Quirk. 1967. 'Adsorption and Desorption of Sulphate Ions by Soil Constituents,' Soil Science, 103:10-15.

Badhe, N. N., and M. G. Lande. 1980. 'Sulphur Supplying Capacity of Different Sulphur-Bearing Compounds as Measured by Its Availability and Uptake by Sorghum (CSH-4) and Wheat (S-227),' Journal of the Maharashtra Agricultural Universities, 5(1):33-35.

Balch, C. C., and G. W. Cooke. 1982. 'The Efficiency of Nutrients and Energy in Plant and Animal Production Systems,' IN Optimizing Yields: The Role of Fertilizers, Proc. 12th Congress, International Potash Institute, Berne, Switzerland.

Bansal, K. N., D. N. Sharma, and D. Singh. 1979. 'Evaluation of Some Soil Test Methods for Measuring Available Sulphur in Alluvial Soils of Madhya Pradesh,' Journal of the Indian Society of Soil Science, 27(3):308-313.

Bansal, K. N., and D. Singh. 1979. 'Nitrogen-Sulphur Ratio for Diagnosing Sulphur Status of Alfalfa,' Journal of the Indian Society of Soil Science, 27(4):452-456.

Barrow, N. J. 1961. 'Studies on the Mineralization of Sulphur from Soil Organic Matter,' Australian Journal of Agricultural Research, 12:306-319.

Beaton, J. D. 1980. 'Sulphur: One Key to High Yields in Small and Coarse Grains,' Solutions, Journal of the National Fertilizer Solutions Association, 24(6):1-8.

Beaton, J. D., D. W. Bixby, S. L. Tisdale, and J. S. Platou. 1974. Fertilizer Sulphur: Status and Potential in the U.S., Tech. Bull. No. 21, The Sulphur Institute, Washington, D.C., U.S.A.

Beaton, J. D., G. R. Burs, and J. Platou. 1968. Determination of Sulphur in Soils and Plant Material, Tech. Bull. No. 14, The Sulphur Institute, Washington, D.C., U.S.A.

Beaton, J. D., and R. L. Fox. 1971. 'Production, Marketing, and Use of Sulphur Products,' IN Fertilizer Technology and Use, 2nd Ed., R. A. Olson et al. (Eds.), pp. 335-379, Soil Science Society of America, Madison, Wisconsin,U.S.A.

Bhan, C., and B. R. Tripathi. 1973. 'The Forms and Contents of Sulphur in Some Soils of U. P.,' Journal of the Indian Society of Soil Science, 21(4):499-504.

Bishop, W. D., and M. S. Mudahar. 1979. 'Fertilizer: Solving the Food Problem for the Developing World,' Ammonia Plant Safety, 21:1-8.

Bixby, D. W. 1979. 'The Outlook for Sulphur,' IN Proceedings of the 29th Annual Meeting Fertilizer Industry Round Table, October 30-November 1, 1979, Washington, D.C., U.S.A., pp. 12-17.

Bixby, D. W. 1980. 'Sulphur Requirements of the Phosphate Fertilizer Industry,' IN The Role of Phosphorus in Agriculture, F. E. Khasawneh, E. C. Sample, and E. J. Kamprath (Eds.), pp. 129-150, American Society of Agronomy, Crop Science Society of America, Soil Science Society of America, Madison, Wisconsin, U.S.A.

Bixby, D. W., and J. D. Beaton. 1970. Sulphur Containing Fertilisers: Properties and Application, Tech. Bull. No. 17, The Sulphur Institute, Washington, D.C., U.S.A.

Bixby, D. W., S. L. Tisdale, and D. L. Rucker. 1964. Adding Plant Nutrient Sulphur to Fertiliser, Tech. Bull. No. 10, The Sulphur Institute, Washington, D.C., U.S.A.

Blair, G. J. 1979. Sulfur in the Tropics, Tech. Bull. IFDC T-12, International Fertilizer Development Center, Muscle Shoals, Alabama, U.S.A., (published jointly by The Sulphur Institute and IFDC).

Blair, G. J., C. P. Mamaril, and E. O. Momuat. 1978. The Sulphur Nutrition of Rice. Contributions, No. 42, Central Research Institute for Agriculture, Bogor, Indonesia.

Blair, G. J., C. P. Mamaril, and E. O. Momuat. 1979. The Sulphur Nutrition of Lowland Rice, IRRI Research Series Paper No. 21, International Rice Research Institute, Los Ba:os, Philippines.

Blair, G. J., C. P. Mamaril, A. Pangerang Umar, E. O. Momuat, and C. Momuat. 1979. 'The Sulphur Nutrition of Rice. 1. A Survey of Soils of South Sulawesi, Indonesia,' Agronomy Journal, 71(3):473-477.

Blair, G. J., E. O. Momuat, and C. P. Mamaril. 1979. 'The Sulphur Nutrition of Rice. 2. Effect of Source and Rate of Sulphur on Growth and Yield Under Flooded Conditions,' Agronomy Journal, 71(3):477-480.

Blair, G. J., and A. J. Nicolson. 1975. 'The Occurrence of Sulfur Deficiency in Temperate Australia,' IN Sulphur in Australasian Agriculture, K. D. McLachlan (Ed.), pp. 137-144, Sydney University Press, Sydney, Australia.

Blair, G. J., P. Paulillian, and S. Samosir. 1978. 'The Effect of Fertilizers on the Yield and Botanical Composition of Pastures in South Sulawesi, Indonesia,' Agronomy Journal, 70(4):559-562.

Blair, G. J., and A. R. Till. 1981. 'Sulphur in Southeast Asia,' Sulphur in Agriculture, 5:5-11.

Blair, G. J., and A. R. Till. 1982. 'Sulphur in Agriculture: Regional Developments in Southeast Asia,' IN Proceedings of the International Sulphur -021-82 Conference, London, November 14-17, 1982, A. I. More (Ed.), Vol. I:477-487, British Sulphur Corporation, Ltd., London, England.

Blair, G. J., and A. R. Till (Eds.). 1983. Sulfur in South-East Asian and South Pacific Agriculture, Proceedings of Research for Dvelopment Seminar, Ciawi, Indonesia, May 23-27, 1983; sponsored by The Australian Development Assistance Bureau (ADAB) and The Sulfur Institute, Washington, D.C., U.S.A.; printed by the University of New England, Armidale, Australia.

Bolle-Jones, E. W. 1964. 'Incidence of Sulphur Deficiency in Africa: A Review,' Empire Journal of Experimental Agriculture, 32(127):241-248.

Bockelee-Morvan, A., and G. Martin. 1966. 'Sulfur Requirements of Peanuts,' Sulphur Institute Journal, 2(2):2-8.

Braud, M. 1970. 'Sulfur Fertilization of Cotton in Tropical Africa,' Sulphur Institute Journal, 5(4):3-5.

British Sulphur Corporation. 1967. World Survey of Ammonium Sulphate, First Edition, The British Sulphur Corporation, Ltd., London, England.

British Sulphur Corporation. 1971. The 1971 World Fertilizer Legislation and Tariffs Manual, London, England.

British Sulphur Corporation. 1972. Ammonium Sulphate in the 1970s, The British Sulphur Corporation, Ltd., London, England.

British Sulphur Corporation. 1974. World Survey of Sulphur Resources, Second Edition, London, England.

British Sulphur Corporation. 1980. 'Company News and Reports,' (original world sulfur consumption estimates by Texasgulf), Sulphur, No. 149, July/August, p. 41.

British Sulphur Corporation. 1981. 'Sulphur Statistics 1979,' (Prepared for ISMA, Ltd.), London, England, A/F/81/87, April.

British Sulphur Corporation. 1982a. 'Preliminary Sulphur and Sulphuric Acid Statistics 1981,'

(Prepared for ISMA, Ltd.), London, England. These figures are also published as 'World Sulphur and Sulphuric Acid Statistics 1981,' Sulphur, No. 161, July/August, pp. 46-48.

British Sulphur Corporation. 1982b. 'World Sulphur and Sulphuric Acid Statistics 1980,' Sulphur, No. 158, January/February, p. 42.

British Sulphur Corporation. 1983a. 'Ammonium Sulphate--Past, Present, and Future,' Nitrogen, No. 141, January-February, pp. 26-30.

British Sulphur Corporation. 1983b. Fertilizer International, various issues, monthly publication of the British Sulphur Corporation, Ltd., London, England.

Bromfield, A. R. 1972. 'Sulphur in Northern Nigerian Soils. (I). Effects of Cultivation and Fertilizers on Total-S and Sulphate Patterns in Soil Profiles,' Journal of Agricultural Science, Cambridge, 78(3):465-470.

Bromfield, A. R. 1973. 'Uptake of Sulfur and Other Nutrients by Groundnuts (Arachis hypogaea L.) in Northern Nigeria,' Experimental Agriculture, 9(1):55-58.

Bromfield, A. R. 1974a. 'The Deposition of Sulphur in Dust in Northern Nigeria,' Journal of Agricultural Science, Cambridge, 83:423-425.

Bromfield, A. R. 1974b. 'The Deposition of Sulphur on Soils in Northern Nigeria,' Journal of Agricultural Science, Cambridge, 83:567-568.

Bromfield, A. R. 1974c. 'The Deposition of Sulphur in Rainwater in Northern Nigeria,' Tellus, 26(3):408-4ll.

Bromfield, A. R. 1974d. 'Report of the Overseas Development Administration Research Project R2122, Sulphur Project in Nigeria,' Overseas Development Administration of the United Kingdom, London, England.

Bromfield, A. R., D. F. Debenham, and I. R. Hancock. 1980. 'The Deposition of Sulphur in Rainwater in Central Kenya,' Journal of Agricultural Science, Cambridge, 94:299-303.

Bromfield, A. R., D. F. Debenham, I. R. Hancock, and M. Powdrill. 1982. – Changes in Soil Sulphur Status and the Development of Sulphur Deficiencies in Tropical Africa,' IN Proceedings of the International Sulphur -021-82 Conference, London, November 14-17, 1982, A. I. More (Ed.), Vol. I:497-515, British Sulphur Corporation, Ltd., London, England.

Bromfield, A. R., I. R. Hancock, and D. F. Debenham. 1982. 'A Collection of Published Papers: Soil and Crop Sulphur Reseach Project R3375, December 1974-March 1980,' Sponsors: Overseas Development Administration of the United Kingdom and Ministry of Agriculture, Kenya.

Brook, R. H. 1979. 'Sulphur in Agriculture,' Abstracts on Tropical Agriculture 5(9):9-20.

Burbano, O. H., and L. M. Blasco. 1975. 'Suelos Volcanos de Nicaragua II, Contentenido y Distribution de Azufre,' Turrialba, 25(4):429-435.

Burki, S. J., and T. J. Goering. 1977. A Perspective on the Foodgrain Situation in the Poorest Countries, World Bank Staff Working Paper No. 251, Washington, D.C., U.S.A.

Burns, G. R. 1967. Oxidation of Sulphur in Soils, Tech. Bull. No. 13, The Sulphur Institute, Washington, D.C., U.S.A.

Calton, W. E., J. W. Vail, and V. P. Padhya. 1961. 'Leaf Composition of Some Tropical Crops,' East African Agricultural and Forestry Journal, 27(1):13-19.

Centro Internacional de Agricultura Tropical (CIAT). 1981. 'Tropical Pastures Program,' CIAT Report 1981: Highlights of Activities in 1980, pp. 63-83, CIAT, Cali, Colombia.

Chahal, R. S., and S. M. Virmani. 1974. 'Uptake and Translocation of Nutrients in Groundnut Plants (Arachis hypogaea L.). III. Sulphur,' Oleagineux, 29(8-9):415-417.

Chao, T. T., M. E. Harward, and S. C. Fang. 1962. 'Adsorption and Desorption Phenomena of Sulphate Ions in Soils,' Soil Science Society of America, Proceedings, 26:234-237.

Chaudhry, I. A., and A. H. Cornfield. 1967. 'Effect of Temperature of Incubation on Sulphate Levels in Aerobic and Sulphide Levels in Anaerobic Soil,' Journal of the Science of Food and Agriculture, 18:82-84.

Child, R. 1957. Manuring of Tea in East Africa, Tea Research Institute of East Africa Pamphlet No. 14, 13 pp.

Chopra, S. L., and J. S. Kanwar. 1966. 'Effect of Sulphur Fertilization on the Chemical Composition and Nutrient Uptake by Legumes,' Journal of the Indian Society of Soil Science, 14(1-4):69-76.

Chopra, S. L., and J. S. Kanwar. 1968. 'Effect of Some Factors on the Transformation of Elemental Sulphur in Soils,' Journal of the Indian Society of Soil Science, 16(1):83-88.

Christensen, R., W. Hendrix, and R. Stevens. 1964. How the United States Improved Its Agriculture, Economic Research Service, USDA, Foreign Agricultural Economic Report No. 76, Washington, D.C., U.S.A.

Coleman, R. 1966. 'The Importance of Sulfur as a Plant Nutrient in World Crop Production,' Soil Science, 101(4):230-239.

Cooper, M. 1968. 'A Comparison of Five Methods for Determining the Sulphur Status of New Zealand Soils,' IN Transactions, International Congress of Soil Science, 9th, Adelaide, 2:263-271.

Council on Environmental Quality and the Department of State. 1981. The Global 2000 Report to the President: Entering the Twenty-First Century, three volumes, Government Printing Office, Washington, D.C., U.S.A.

Couto, W., D. J. Lathwell, and D. R. Bouldin. 1979. 'Sulphate Sorption by Two Oxisols and an Alfisol of the Tropics,' Soil Science, 127(2):108-116.

Crosson, P. R., and K. D. Frederick. 1977. The World Food Situation: Resource and Environmental Issues in the Developing Countries and the United States, Research Paper R-6, Resources for the Future, Washington, D.C., U.S.A.

Dabin, B. 1972. "Preliminary Results of a Survey Showing the Sulphur Content of the Soils of Tropical Africa," IN Proceedings of International Symposium on Sulphur in Agriculture, Versailles, 3-4.XII.1970, nnales Agronomiques, Num:ro hors s:rie, pp. 113-136.

Daigger, L. A., and R. L. Fox. 1971. 'Nitrogen and Sulphur Nutrition of Sweet Corn in Relation to Fertilization and Water Composition,' Agronomy Journal, 63:729-730.

Dalal, J. L., J. S. Kanwar, and J. S. Saini. 1963. 'Investigations on Soil Sulphur. II. Gypsum as a Fertilizer for Groundnut in the Punjab,' Indian Journal of Agricultural Sciences, 33(3):199-204.

Das, S. K., P. Chhabra, S. R. Chatterjee, Y. P. Abrol, and D. L. Deb. 1975. 'Influence of Sulphur Fertilisation on Yield of Maize and Protein Quality of Cereals,' Fertiliser News, 20:30-32.

Decau, J. 1972. "Sulphur Supplied by Organic Manures," IN Proceedings of International Symposium on Sulphur in Agriculture, Versailles, 3-4.XII.1970, Annales Agronomiques, Num:ro hors s:rie, pp. 315-322.

De Freitas, L.M.M., A. C. McClung, and W. L. Lott. 1960. Field Studies on Fertility Problems of Two Brazilian Campos Cerrados 1958-59, IRI Bulletin 21, IRI Research Institute, Inc., New York, New York, U.S.A.

De Freitas, L.M.M., F. P. Gomes, and W. L. Lott. 1972. 'Effects of Sulfur Fertiliser on Coffee,' Sulphur Institute Journal, 8(3):9-12 (also available as IRI Bulletin 41, IRI Research Institute, Inc., New York, New York, U.S.A.).

De Silva, M.A.T., G. M. Anthonypillai, and D. T. Mathes. 1977. 'Sulphur Nutrition of Coconut,' Experimental Agriculture, 13(3):265-271.

Dev, G., R. C. Jaggi, and M. S. Aulakh. 1979. 'Study of Nitrate-Sulphate Interaction on the Growth and Nutrient Uptake by Maize Using ^{35}S,' Journal of the Indian Society of Soil Science, 27(3):302-307.

Dev, G., and V. Kumar. 1982. 'Secondary Nutrients,' IN Review of Soil Research in India, Part I, pp. 342-360, 12th International Congress of Soil Science, New Delhi, India.

Dev, G., and S. Saggar. 1974. 'Effect of Sulfur Fertilization on the N:S Ratio in Soybean Varieties,' Agronomy Journal, 66:454-455.

Dev, G., S. Saggar, and M. S. Bajwa. 1981. 'Nitrogen-Sulphur Relationships in Raya (Brassica juncea L.) Cultivars as Influenced by Sulphur Fertilization,' Journal of the Indian Society of Soil Science, 29(3):397-399.

228

Dhillon, N. S., and G. Dev. 1978. 'Effect of Elemental Sulfur Application on the Soybean (Glycine max. Merril), Journal of the Indian Society of Soil Science, 26(1):55-57.

Dijkshoorn, W., and A. L. van Wijk. 1967. 'The Sulphur Requirements of Plants as Evidenced by the Sulphur-Nitrogen Ratio in the Organic Matter: A Review of Published Data,' Plant and Soil, 26:129-157.

Do Nascimento, J.A.L., and M. Morelli. 1980. 'Sulphur in Soils of Rio Grand do Sul: Forms in the Soil,' Revista Brasillera de Ciencia Solo Brazil, 4(3):131-135.

Dube, S. D., and P. H. Misra. 1970. 'Effect of Sulphur Deficiency on Growth, Yield, and Quality in Some of the Important Leguminous Crops,' Journal of the Indian Society of Soil Science, 18(4):375-378.

Dudal, R. 1980. "Soil-Related Constraints to Agricultural Development in the Tropics," IN Priorities for Alleviating Soil-Related Constraints to Food Production in the Tropics, pp. 23-37, jointly sponsored and published by the Inernational Rice Research Institute and the New York State College of Agriculture and Life Sciences, Cornell University, in cooperation with the University Consortium of Soils for the Tropics, IRRI, Los Ba:os, Philippines.

Dutt, A. K. 1962a. 'Sulphur Deficiency in Sugarcane,' Empire Journal of Experimental Agriculture, 30:119.

Dutt, A. K. 1962b. 'Sulphur Deficiency of Jute,' Tropical Agriculture (Trinidad), 39(1):73-76.

Ensminger, L. E. 1954. 'Some Factors Affecting the Adsorption of Sulphate by Alabama Soils,' Soil Science Society of America, Proceedings, 18:259-264.

Ensminger, L. E. 1958. Sulphur in Relation to Soil Fertility, Alabama Agricultural Experiment Station Bulletin No. 312, Auburn, Alabama, U.S.A.

Enwezor, W. O. 1976. 'Sulfur Deficiency in Soils of Southeastern·Nigeria,' Geoderma, 15:401-411.

Eppendorfer, W. H. 1971. 'Effects of Sulfur, Nitrogen, and Phosphorus on Amino Acid Composition of Field Beans (Vicia faba) and Responses of the Biological Value of the Seed Protein and Sulfur-Amino Acid Content,' Journal of the Science of Food and Agriculture, 22(10):501-505.

Ferguson, H., N. G. Gokhale, and S. K. Dutta. 1957. 'Urea and Ammonium Sulphate-Nitrate,' Newsletter Indian Tea Assoc., Tocklai, 4:3.

Fertiliser Association of India (FAI). 1983. Fertiliser Statistics (and various previous issues), Annual publication by Fertiliser Association of India, New Delhi, India.

Fitts, J. W. 1970. 'Sulphur Deficiency in Latin America,' Sulphur Institute Journal, 6(2):14-16.

FAO/UNIDO/World Bank. 1983. 'Current World Fertilizer Situation and Outlook, 1981/82-1987/88,' (and various previous issues), FAO/UNIDO/World Bank Working Group, Food and Agriculture Organization of the United Nations (FAO), Rome, Italy.

Food and Agriculture Organization of the United Nations (FAO). 1977. The Fourth World Food Survey, Food and Nutrition Series No. 10, FAO, Rome, Italy.

Food and Agriculture Organization of the United Nations (FAO). 1979. Agriculture: Toward 2000, FAO Conference, Twentieth Session, Nov. 10-29, FAO, Rome, Italy.

Food and Agriculture Organization of the United Nations (FAO). 1982. Production Yearbook, 1981 (and various previous issues and the FAO production data tape), FAO, Rome, Italy.

Food and Agriculture Organization of the United Nations (FAO). 1983. Fertilizer Yearbook, 1982 (and various previous issues), FAO, Rome, Italy.

Food and Agriculture Organization of the United Nations (FAO). 1984. Monthly Bulletin of Statistics, Vol. 7, March 1984, FAO, Rome, Italy.

Fox, R. L. 1980. 'Responses to Sulphur by Crops Growing in Highly Weathered Soils,' Sulphur in Agriculture, 4:16-22.

Fox, R. L., B. T. Kang, and D. Nangju. 1977. 'Sulfur Requirements of Cowpea and Implications for Production in the Tropics,' Agronomy Journal, 69(2):201-205.

Fox, R. L., D. G. Moore, J. M. Wang, D. L. Pluncknett, and R. D. Furr. 1965. −Sulphur in Soils, Rainwater, and Forage Plants of Hawaii,' Hawaii Farm Science, 14(3):9-12.

Frederick, M. T. 1983. 'Review of Alternatives and Recommendations for Using Phosphogypsum as an Agricultural Sulfur Source for Bangladesh,' Report submitted to Bangladesh Agricultural Development Corporation, International Fertilizer Development Center, Muscle Shoals, Alabama, U.S.A.

Free, W. J., B. J. Bon, and J. L. Nevins. 1976. Changing Patterns in Agriculture and Their Effect on Fertilizer Use, Bulletin Y-106, TVA, Muscle Shoals, Alabama, U.S.A.

Freney, J. R., G. E. Melville, and C. H. Williams. 1970. 'The Determination of Carbon Bonded Sulfur in Soil,' Soil Science, 109:310-318.

Freney, J. R., P. J. Randall, and K. Spencer. 1982. 'Diagnosis of Sulphur Deficiency in Plants,' IN Proceedings of the International Sulphur '82 Conference, held in London, November 14-17, 1982, A. I. More (Ed.), Vol. I:439-444, British Sulphur Corporation, Ltd., London, England.

Freney, J. R., F. J. Stevenson, and A. H. Beavers. 1972. 'Sulphur-Containing Amino Acids in Soil Hydrolysates,' Soil Science, 114:468-476.

Fritz, A. 1972. 'Sulphur and Its Role as Plant Nutrient in Tropical and Subtropical Agriculture,' Agricultural Bulletin 1E, Badische Anilin ' SodaFabrik AG, Ludwigshafen, Federal Republic of Germany.

Galasch, H. E. 1976. 'Coconut Nutrition in the Markham Valley of New Guinea,' Papua and New Guinea Agricultural Journal, 27(4):75-91.

Ghosh, A. B. 1980. 'Sulphur in Relation to Soil and Crop Situations in India,' Fertiliser News, 25(12):36-39.

Goldsworthy, P. R., and R. G. Heathcote. 1963. 'Fertilizer Trials With Groundnuts in Northern Nigeria,' Empire Journal of Experimental Agriculture, 31:351-366.

Gosnell, J. M. 1964. 'A Sulphur Deficiency in Wattle,' East African Agricultural and Forestry Journal, 30:1-7.

Gosnell, J. M., and A. C. Long. 1969. 'A Sulphur Deficiency in Sugarcane,' Proceedings of the South African Sugar Technologists Association, June.

Grant, P. M., E. W. Hughes, J. Moerman, and D. T. Coker. 1964. Soil Chemistry, Annual Report of Agricultural Research Council Cent. Agr. 1964, p. 15-21.

Grant, P. M., and A.W.G. Rowell. 1976. 'Studies on Sulphate Fertilisers for Rhodesian Crops. 1. Effect of Sulphur in Fertiliser Compounds on the Yield and Sulphur Status of Maize,' Rhodesian Journal of Agricultural Research, 14(2):101-109.

Grant, P. M., and A.W.G. Rowell. 1978. 'The Distribution of Sulphate and Total Sulphur in Maize Plants (Zea mays L.) in Relation to the Diagnosis of Deficiency,' Rhodesian Journal of Agricultural Research, 16(1):43-59.

Grant, P. M., and T. F. Shaxton. 1970. 'The Effect of Ammonium Sulphate Fertilizer on the Sulphur Content of Tea Garden Soils in Malawi,' Tropical Agriculture (Trinidad), 47(1):31-36.

Green Markets: Fertilizer Market Intelligence Weekly. 1984. 7(5) (various previous issues), McGraw-Hill, Inc., New York, New York, U.S.A.

Greenwood, M. 1951. 'Fertilizer Trials with Groundnuts in Northern Nigeria,' Empire Journal of Experimental Agriculture, 19:225-244.

Guerarao, R. R., and H. B. Orjuela. 1979. 'Fracciones de Azufre y Niveles Criticos de Disponibilidad Para la Planta en Suelos de los Llanos Orientales y la Sabana de Bogota,' Suelos Ecuatoriales, 10(2):232-244.

Gupta, J. K., and Y. P. Gupta. 1972. 'Note on Effect of Sulphur and Molybdenum on Quality of Pea,' Indian Journal of Agronomy, 17(3):245-247.

Gupta, V. K., and D. S. Mehta. 1980. 'Influence of Sulphur on the Yield and Concentration of Copper, Manganese, Iron, and Molybdenum in Berseem (Trifolium alexandrinum) Grown on Two Different Soils,' Plant and Soil, 56(2):229-234.

Halais, P., and A. Girault. 1973. 'Correction of Sulphur Deficiency in Sugarcane in Madagascar,' Revue Agricole et Sucriere de L-021-Ile Maurice, 52:244-250.

Haldhar, B. R., and H. P. Barthakur. 1976. 'Sulphide Production in Some Typical Rice Soils of Assam,' Journal of the Indian Society of Soil Science, 24(4):387-395.

230

Haque, I. 1971. 'The Status of Sulphurin West Indian Soils,' Ph.D. diss., The University of the West Indies, St. Augustine, Trinidad.

Haque, I., and D. Walmsley. 1974. 'Sulphur Investigation in Some West Indian Soils,' Tropical Agriculture (Trinidad), 51(2):253-263.

Harward, M. E., T. T. Chao, and S. C. Fang. 1962. 'The Sulfur Status and Sulfur Supplying Power of Oregon Soils,' Agronomy Journal, 54:101-106.

Harward, M. E., and H. M. Reisenauer. 1966. 'Reactions and Movement of Inorganic Soil Sulfur,' Soil Science, 101:326-335.

Hasan, S. M., R. L. Fox, and C. C. Boyd. 1970. 'Solubility and Availability of Sorbed Sulfate in Hawaiian Soils,' Soil Science Society of America, Proceedings, 34:897-901.

Hayami, Y., and V. W. Ruttan. 1971. Agricultural Development: An International Perspective, The Johns Hopkins University Press, Baltimore, Maryland, U.S.A.

Hazleton, J. E. 1970. The Economics of the Sulphur Industry, Resources for the Future, Inc., Washington, D.C., U.S.A.

Hendrix, J. E. 1967. 'The Effect of pH on the Uptake and Accumulation of Phosphate and Sulfate Ions by Bean Plants,' American Journal of Botany, 54(5), Pt. 1:560-564.

Herdt, R. W., and R. Barker. 1975. 'Possible Effects of Fertilizer Shortages on Rice Production in Asian Countries,' IN Impact of Fertilizer Shortage: Focus on Asia, Asian Productivity Organization, Tokyo, Japan.

Hignett, T. P. 1970. 'Trends in Fertilizer Materials 1970-1980,' Proceedings of the 20th Annual Meeting Fertilizer Industry Round Table, pp. 1-7.

Hignett, T. P. 1974. 'Recent Developments in Fertilizer Production Technology and Economics With Special Reference to Ammonia and Compound Fertilizers,' Extension Bulletin 40, Asian and Pacific Food and Fertilizer Technology Center, Taipei, Taiwan.

Hignett, T. P. 1979. 'Technology of Production of Fertilizers Containing Ca, Mg, and S,' Paper presented at the Colombian Soil Science Colloquium on Saline Soils and Secondary Elements in Colombian Agriculture, September 1921, Palmira, Colombia.

Hignett, T. P., and P. J. Stangel. 1982. 'Agricultural Sulfur in the Americas,' IN Proceedings of the International Sulfur -021-82 Conference, held in London, November 14-17, 1982, A. I. More (Ed.), Vol. I:465-475, British Sulphur Corporation, Ltd., London, England.

Hill, C.C.H. 1963. Kenya Department of Agricultural Annual Report, 11:116-118.

Hira, G. S., and N. T. Singh. 1980. 'Irrigation Water Requirements for Dissolution of Gypsum in Sodic Soils,' Soil Science Society of America Journal, 44:930-933.

Hoeft, R. G. 1981. "Crop Response to Sulphur," El Uso del Azufre para el Desarrollo y Modernización de la Agricultura en América Latina, 1.er Simposia, México, D.F., México.

Hoeft, R. G., L. M. Walsh, and D. R. Keeney. 1973. 'Evaluation of Various Extractants for Available Soil Sulphur,' Soil Science Society of America, Proceedings, 37:401-404.

Hopper, W. D. 1981. 'Recent Trends in World Food and Population,' IN Future Dimensions of World Food and Population, R. G. Woods (Ed.), pp. 35-55, Westview Press, Boulder, Colorado, U.S.A.

Hoque, M. Z., and P. R. Hobbs. 1980. 'Response of Rice Crops to Added Sulphur at BRRI Station and Nearby Project Area,' IN Workshop on Sulphur Nutrition in Rice, December 1978, pp. 15-19, Publication No. 41, Bangladesh Rice Research Institute, Joydebpur, Dhaka, Bangladesh.

Hoque, M. Z., and M.A.H Khan. 1980. 'Potential for Increasing the Yields of Farmers' Aus Rice in Bangladesh,' International Rice Commission Newsletter, 29(1):39-42.

Horseman, M.N.J. 1973. World Sulphur Supply and Demand: 1960-1980 (Report prepared for United Nations Industrial Development Organization), United Nations, New York, New York, U.S.A.

Huacuja, R. F., and L. J. Cajuste. 1981. "Fijacion de Azufre en Suelos Derivados de Cenizas Volcanicas de la Sierra Tarasca," El Uso del Azufrepara el Desarrollo y Modernización de la Agricultura en América Latina, 1.er Simposio, México, D.F., México.

International Fertilizer Development Center (IFDC). 1979. Fertilizer Manual, IFDC-R-1, (also available from United Nations Industrial Development Organization, Vienna, Austria), Muscle Shoals, Alabama, U.S.A.

International Food Policy Research Institute (IFPRI). 1977. Food Needs of Developing Countries: Projections of Production and Consumption to 1990, Research Report 3, Washington, D.C., U.S.A.

International Petroleum Encyclopedia. 1982. The Petroleum Publishing Company, Tulsa, Oklahoma, U.S.A. (Also December 28, 1981, issue of Oil and Gas Journal).

Islam, M. M., and F. N. Ponnamperuma. 1982. 'Soil and Plant Tests for Available Sulfur in Wetland Rice Soils,' Plant and Soil, 68:97-113.

ISMA. 1982. 'Fertilizer Consumption Statistics 1980/81,' (and various previous annual issues), Paris, France (ISMA, International Phosphate Industry Association, Ltd., has changed to IFA, International Fertilizer Industry Association, Ltd.).

Ismunadji, M., and M. Miyake. 1978. 'Sulphur Application and Amino Acid Content of Brown Rice,' Japan Agricultural Research Quarterly, 12(3):180-182.

Ismunadji, M., and I. Zulkarnaini. 1978. 'Sulphur Deficiency of Lowland Rice in Indonesia,' Sulphur in Agriculture, 2:17-19,22.

Ismunadji, M., I. Zulkarnaini, and M. Miyake. 1975. Sulfur Deficiency in Lowland Rice in Java. Contributions, No. 14, Central Research Institute for Agriculture, Bogor, Indonesia.

Jacob, K. D. 1964. 'History and Status of the Superphosphate Industry,' IN Superphosphate: Its History, Chemistry and Manufacture, pp. 19-94, U.S. Department of Agriculture and Tennessee Valley Authority, Government Printing Office, Washington, D.C., U.S.A.

Jaggi, T. N. 1982. 'Amjhore Pyrites as a Source of Agricultural Sulphur and an Amendment for Alkaline Soils,' IN Proceedings of the International Sulphur '82 Conference, held in London, November 14-17, 1982, A. I. More (Ed.), Vol. I:547-560, British Sulphur Corporation, Ltd., London, England.

Johnson, C. M., and H. Nishita. 1952. 'Microestimation of Sulphur in Plant Materials, Soils, and Irrigation Waters,' Analytical Chemistry, 24:736-742.

Johnston, B. F., and P. Kilby. 1975. Agriculture and Structural Transformation, Oxford University Press, New York, New York, U.S.A.

Johnston, B. F., and J. W. Mellor. 1961. 'The Role of Agriculture in Economic Development,' American Economic Review, 51:566-593.

Jones, M. J. 1977. Sulphur in Malawi Soils: A Review, Research Bulletin, 8:1-12, Bunda College Agriculture, University of Malawi.

Jones, U. S. 1978. 'Sulphur Content of Rain Water in South Carolina,' IN Environmental Chemistry and Cycling Processes, D. C. Adriano and I. L. Brisbin, Jr. (Eds.), U.S. Department of Energy, Symposium Sries 45, pp. 394-402.

Jordan, H. V., and H. M. Reisenauer. 1957. 'Sulphur and Soil Fertility,' IN Soil: The Yearbook of Agriculture, pp. 107-111, U.S. Department of Agriculture (USDA), U.S. Government Printing Office, Washington, D.C., U.S.A.

Joshi, D. C., J. S. Choudhari, and S. V. Jain. 1973. 'Distribution of Sulphur Fractions in Relation to Forms of Phosphorus in Soils of Rajasthan,' Journal of the Indian Society of Soil Science, 21(3):289-294.

Joshi, D. C., and S. P. Seth. 1975. 'Effect of Sulphur and Phosphorus Application on Soil Characteristics, Nutrient Uptake, and Yield of Wheat Crop,' Journal of the Indian Society of Soil Science, 23(2):217-221.

Kamarck, A. M. 1976. The Tropics and Economic Development: A Provocative Inquiry Into the Poverty of Nations, The Johns Hopkins University Press, Baltimore, Maryland, U.S.A.

Kampfer, M., and E. Zehler. 1967. "The Importance of the Sulphate Fertilizers for Raising the Yield and Improving the Quality of Agricultural, Horticultural, and Silvicultural Crops," Potash Review 24'28, Berne, Switzerland.

Kamprath, E. J. 1972. 'Sulphur,' IN Review of Soils Research in Tropical Latin America, P. A. Sanchez (Ed.), pp. 238-243, North Carolina State University, Raleigh, North Carolina, U.S.A.

Kamprath, E. J. 1981. "Agricultural Requirements for Sulfur in South and Central America," El Uso del Azufre para el Desarrollo y Modernización de la Agricultura en América Latina, 1.er Simposio, México, D.F., México.

Kamprath, E. J., and U. S. Jones. 'Plant Response to Sulfur in the Southeastern United States,' IN Sulphur in Agriculture, M. A. Tabatabai (Ed.), American Society of Agronomy (In press).

Kamprath, E. J., W. L. Nelson, and J. W. Fitts. 1956. 'The Effect of pH, Sulphate, and Phosphate Concentrations on the Adsorption of Sulphate by Soils,' Soil Science Society of America, Proceedings, 20:463-466.

Kamprath, E. J., W. L. Nelson, and J. W. Fitts. 1957. 'Sulfur Removed From Soils by Field Crops,' Agronomy Journal, 49:289-293.

Kang, B. T., E. Okoro, D. Acquaye, and O. A. Osiname. 1981. 'Sulfur Status of Some Nigerian Soils From the Savanna and Forest Zones,' Soil Science, 132(3):220-227.

Kang, B. T., and O. A. Osiname. 1976. 'Sulfur Response of Maize in Western Nigeria,' Agronomy Journal, 68(2):333-336.

Kanwar, J. S. 1963. 'Investigations on Sulphur in Soils. 1. Sulphur Deficiency in Groundnut Soils of Samrala (Ludhiana),' Indian Journal of Agricultural Science, 33(1):196-198.

Kanwar, J. S. 1982. 'Managing Soil Resources to Meet the Challenges to Mankind,' Presidential Address, 12th International Congress of Soil Science, Feb. 8-16, New Delhi, India.

Kanwar, J. S., and S. Mohan. 1964. 'Sulphur in Soils: Distribution of Forms of Sulphur in Punjab Soils,' Presented at the Symposium on Fertility of Indian Soils, held on August 3-4, 1962. Reprinted from the Bulletin of the National Institute of Sciences of India, No. 26:31-36.

Kanwar, J. S., H. L. Nijhawan, and S. K. Raheja. 1983. Groundnuts: Its Nutrition and Fertilizer Responses in India, p. 132, Indian Council of Agricultural Research, New Delhi, India.

Kanwar, J. S., and N. S. Randhawa. 1974. Micronutrient Research in Soils and Plants in India: A Review, ICAR Technical Bulletin (Agric.) No. 50:147-158, Second Edition, Indian Council of Agricultural Research, New Delhi, India.

Kanwar, J. S., and P. N. Takkar. 1964. 'Distribution of Sulfur Forms in Tea Soils of the Punjab,' Journal of Research, Punjab Agricultural University, Ludhiana, 1(1):1-15.

Kanwar, J. S., and P. N. Takkar. 1966. 'Responses to Sulphur in Tea Soils of the Punjab,' Journal of Research, Punjab Agricultural University, Ludhiana, 3(3):246-252.

Karim, A., S. M. Alam, and M. Rahman. 1970. Annual Technical Report, Atomic Energy Centre, Dhaka, Bangladesh.

Karim, M., and M.A.K. Majlish. 1958. 'A Study of the Formative Effects of Sulphur on Rice Plant,' Pakistan Journal of Scientific Research, 10:52.

Keerati-Kasikorn, P. 1982. 'Sulphur Deficiency in Northeast Thailand,' IN Proceedings of the International Sulphur '82 Conference, held in London, November 14-17, 1982, A. I. More (Ed.), Vol. I: British Sulphur Corporation, Ltd., London, England.

Kilmer, V. J., and D. C. Nearpass. 1960. 'The Determination of Available Sulphur in Soils,' Soil Science Society of America, Proceedings, 24:337-340.

Kiyoura, R. 1982. 'Sulphur Dioxide Should Be Reconsidered As Sulphur Source Necessary for Preservation of Life and Vitality: Excessive Restrictions are Unnecessary and Harmful,' Chemical Economy and Engineering Review, 14(3):23-30.

Korentajer, L., B. H. Byrnes, and D. T. Hellums. 1983. 'The Effect of Liming and Leaching on Sulphur-Supplying Capacity of Soils,' Soil Science Society of America Journal, 47(3):525-530.

Korentajer, L., U. Mokwunye, and D. T. Hellums. 1982. 'Greenhouse Evaluation of Phosphorus Fertilizer Materials Containing Sulfur,' draft, International Fertilizer Development Center (IFDC), Muscle Shoals, Alabama, U.S.A.

Kumar, V., and M. Singh. 1979. 'Sulphur and Zinc Relationship on Uptake and Utilization of Zinc in Soybean,' Soil Science, 128(6):343-347.

Kumar, V., and M. Singh. 1980. 'Sulphur, Phosphorus, and Molybdenum Interactions in Relation to Growth, Uptake, and Utilization of Sulphur in Soybean,' Soil Science, 129(5):297-304.

Kumar, V., M. Singh, and N. Singh. 1981. 'Effect of Sulphate, Phosphate, and Molybdate Application on Quality of Soybean Grain,' Plant and Soil, 59:3-8.

Laurence, R.C.N., R. W. Gibbons, and C. T. Young. 1976. 'Changes in the Yield, Protein, Oil, and Maturity of Groundnut Cultivars With the Application of Sulphur Fertilizers and Fungicides,' Journal of Agricultural Science, Cambridge, 86(2):245-250.

Leggett, J. E., and E. Epstein. 1956. 'Kinetics of Sulphate Adsorption by Barley Roots,' Plant Physiology, 31:222-226.

Lockard, R. G., J. O. Ballaux, and E. A. Liongson. 1972. 'Response of Rice Plants Grown in Three Potted Luzon Soils to Additions of Boron, Sulfur, and Zinc,' Agronomy Journal, 64:444-447.

Lott, W. L., A. C. McClung, and J. C. Medcalf. 1960. Sulphur Deficiency in Coffee, IRI Bulletin 22, IRI Research Institute, Inc., New York, New York, U.S.A.

Malavolta, E. 1952. 'Chemical and Agricultural Studies Pertaining to Sulphur,' Anais da Escola Superior de Agricultura Luiz de Queroz, Vol. 9, pp. 40-130, Piracicaba, Sao Paulo, Brazil.

Malavolta, E. 1979. "Potássio, Magnésio E Enxofre Nos Solos e Culturas Brasileiras," Tech. Bull. No. 4, Potash and Phosphate Institute, Atlanta, Georgia, U.S.A., and the International Potash Institute, Berne, Switzerland.

Mamaril, C. P., A. Pangerang Umar, I. Manwan, and C.J.S. Momuat. 1976. 'Sulfur Response of Lowland Rice in South Sulawesi, Indonesia,' Contributions, No. 22, Central Research Institute for Agriculture, Bogor, Java, Indonesia.

Manuel Arrando, M., A. Chueca Sancho, M. Gomez Ortega, J. Lopez George, and L. Recalde Martinez. 1976. 'Effect of Elemental Sulphur Application and Nitrogen Fertilization on the Yield and Protein Content of Wheat,' Anales de Edafologia y Agrobiologia, 35(7-8):781-795.

—Markets Newsletter.— 1982. Chemical Week, December 1, p. 68.

Marok, A. S. 1978. 'Sulphur Deficiency Limits Wheat Yields in Arid Brown Soil,' Indian Journal of Agronomy, 23(2):268-269.

Marok, A. S., and G. Dev. 1979. 'Responses of Sunflower (Herianthus annus) to Sulphur Application and Evaluation of Sulfur Status of Soils,' Journal of Nuclear Agricultural Biology, 8(3):100-102.

Marok, A. S., and G. Dev. 1980a. 'Phosphorus and Sulphur Interrelationships in Wheat,' Journal of the Indian Society of Soil Science, 28(2):184-188.

Marok, A. S., and G. Dev. 1980b. 'Phosphorus Sulfur Relationships in Berseem (Trifolium alexandrinum) as Measured by Yield and Plant Analysis,' Journal of Nuclear Agricultural Biology, 9(2):54-56.

Martin-Pr:vel, P. 1972. "Sulphur Needs and Uptake by Various Tropical Crops," IN Proceedings of International Symposium on Sulphur in Agriculture, Versailles,

3-4.XII.1970, Annales Agronomiques, Num:ro hors s:rie, pp. 8l-99.

Masters, M., and R. A. McCance. 1939. 'The Sulphur Content of Foods,' Biochemical Journal, 33:1304-1312.

McClung, A. C., L.M.M. de Freitas, and W. L. Lott. 1959. 'Analyses of Several Brazilian Soils in Relation to Plant Responses to Sulphur,' Soil Science Society of America, Proceedings, 23:221-229.

McClung, A. C., L.M.M. de Freitas, D. S. Mikkelsen, and W. L. Lott. 1962. Cotton Fertilisation in Campo Cerrado Soils of State of Sao Paulo, Brazil, IRI Bulletin 27, IRI Research Institute, Inc., New York, New York, U.S.A.

McCune, D. L. 1982. Fertilizers for Tropical and Subtropical Agriculture, (Twelfth Francis New Memorial Lecture, The Fertiliser Society, London, March 12, 1982). Special Publication SP-2, International Fertilizer Development Center, Muscle Shoals, Alabama, U.S.A.

McNaught, K. J., and C. During. 1970. 'Relations Between Nutrient Concentrations in Plant

234

Tissues and Responses of White Clover to Fertilizers on a Gley Podzol Near Westport,' New Zealand Journal of Agricultural Research, 13:567-590.

Mehlich, A. 1970. 'Crop Responses to Sulphur in Kenya,' Sulphur Institute Journal, 5(4):10-13.

Mehta, U. R., and H. G. Singh. 1979. 'Response of Greengram to Sulphur on Calcareous Soils,' Indian Journal of Agricultural Sciences, 49(9):703-706.

Mellor, J. W. 1965. The Economics of Agricultural Development, Cornell University Press, Ithaca, New York, U.S.A.

Mellor, J. W. 1976. The New Economics of Growth: A Strategy for India and the Developing World, Cornell University Press, Ithaca, New York, U.S.A.

Mengel, K. M., and E. R. Kirkby. 1982. Principles of Plant Nutrition (3rd Edition), International Potash Institute, Berne, Switzerland.

Messing, J.H.L. 1970. 'In the West Indies Bananas Respond to Sulphur,' Sulphur Institute Journal, 5(4):6-7.

Metson, A. J. 1973. Sulphur in Forage Crops, Tech. Bull. No. 20, The Sulphur Institute, Washington, D.C., U.S.A.

Metson, A. J., and L. C. Blakemore. 1978. 'Sulphate Retention by New Zealand Soils in Relation to the Competitive Effect of Phosphate,' New Zealand Journal of Agricultural Research, 21:243-253.

Meyer, B. 1977. Sulfur, Energy and Environment, Elsevier Scientific Publishing Company, Amsterdam, The Netherlands.

Mikkelsen, D. S., L.M.M. de Freitas, and A. C. McClung. 1963. Effects of Liming and Fertilizing Cotton, Corn, and Soybeans on Campo Cerrado Soils--State of Sao Paulo, Brazil, IRI Bulletin 29, IRI Research Institute, Inc., New York, New York, U.S.A.

Mitchel, C. C., Jr., and W. G. Blue. 1981. 'The Sulfur Fertility Status of Florida Soils. II. An Evaluation of Subsoil Sulfur on Plant Nutrition,' Proceedings of the Soil and Crop Science Society, Florida, 40:77-82.

Mudahar, M. S. 1978. 'Needed Information and Economic Analysis for Fertilizer Policy Formulation,' Indian Journal of Agricultural Economics, 33(3), July/September (also available as ADC Teaching and Research Forum Reprint No. 24, September 1980).

Mudahar, M. S., and T. P. Hignett. 1982. Energy and Fertilizer: Policy Implications and Options for Developing Countries, Tech. Bull. IFDC-T-20, International Fertilizer Development Center, Muscle Shoals, Alabama, U.S.A.

Munson, R. D. 1982. Potassium, Calcium, and Magnesium in the Tropics and Subtropics, Tech. Bull. IFDC-T-23, International Fertilizer Development Center, Muscle Shoals, Alabama, U.S.A.

Murphy, M. D., and J. C. Brogan. 1981. "Predicting Sulfur Deficiency by Soil and Plant Analysis," El Uso del Azufre para Desarrollo Modernización de la Agricultura en America Latina, 1.er Simposio, México, D.F., México.

Naik, M. S., and N. B. Das. 1964. 'Available Sulphur Status of Indian Soils by the Aspergillus Niger Method,' Journal of the Indian Society of Soil Science, 12(3):151-155.

National Fertilizer Development Center (NFDC). 1982. World Fertilizer Market Information Service, Tennessee Valley Authority, Muscle Shoals, Alabama, U.S.A.

Nearpass, D. C., M. Fried, and V. J. Kilmer. 1961. 'Greenhouse Measurement of Available Sulphur Using Radioactive Sulphur,' Soil Science Society of America, Proceedings, 25(4):287-289.

Neller, J. R. 1959. 'Extractable Sulphate Sulfur in Soils of Florida in Relation to the Amount of Clay in the Profile,' Soil Science Society of America, Proceedings, 23:346-348.

Neptune, A.M.L., M. A. Tabatabai, and J. J. Hanway. 1975. 'Sulfur Fractions and Carbon-Nitrogen-Phosphorus-Sulfur Relationships in Some Brazilian and Iowa Soils,' Soil Science Society of America, Proceedings, 39:51-55.

Ngongi, A.G.N., R. Howeler, and H. A. MacDonald. 1977. 'Effect of Potassium and Sulphur on Growth, Yield, and Composition of Cassava,' IN Proceedings of the 4th Symposium of the International Society for Tropical Root Crops, pp. 107-113.

Noggle, J. C. 1980. 'Sulfur Accumulation by Plants; the Role of Gaseous Sulfur in Crop Nutrition,' IN Atmospheric Sulphur Deposition: Environmental Impact and Health Effects, D. S. Shriner, C. R. Richmond, and S. E. Lindberg, (Eds.), pp. 289-297, Proceedings of the Second Life Sciences Symposium, Potential Environmental and Health Consequences of Atmospheric Sulphur Deposition, Gatlinburg, Tennessee, October 14-18, 1979, Ann Arbor Science Publishers, Inc., Ann Arbor, Michigan, U.S.A.

No, Y. M. 1981. 'Sulphur Mineralization and Adsorption in Soils,' Plant and Soil, 60(3):451-459.

O'Connor, K. F., and E. W. Vartha. 1969. 'Responses of Grasses to Sulphur Fertilizers,' New Zealand Journal of Agricultural Research, 12:97-118.

Oke, O. L. 1967. 'Sulphur Status of Some Nigerian Soils,' Journal of the Indian Society of Soil Science, 15(1-4):207-208.

Oke, O. L. 1969. 'Sulphur Nutrition of Legumes,' Experimental Agriculture, 5(2):111-116.

Oke, O. L. 1970. 'Sulphur Status of Nigerian Soils and Uptake by Grasses,' Journal of the Indian Society of Soil Science, 18(2):163-169.

Pareek, S. K., M. S. Saroha, and H. G. Singh. 1978. 'Effect of Sulphur on Chlorosis and Yield of Blackgram on Calcareous Soils,' Indian Journal of Agronomy, 23(3):102-107.

Pasricha, N. S., M. S. Bajwa, and N. S. Randhawa. 1975. 'Sulphur for LightTextured Soils,' Indian Farming, 25(2):9,28.

Pasricha, N. S., and N. S. Randhawa. 1973. 'Sulphur Nutrition of Crops From Native and Applied Sources,' Indian Journal of Agricultural Sciences, 43(3):270-274.

Pasricha, N. S., and N. S. Randhawa. 1975. 'Effect of Sulphur Fertilization on the Nitrogen Metabolism in Berseem,' Indian Journal of Agricultural Sciences, 45(5):213-218.

Pasricha, N. S., N. S. Randhawa, G. S. Bahl, and G. Dev. 1977. 'Sulphur Uptake and Dry-Matter Production in Maize at Different Growth Stages as Affected by Native and Applied Sulphur,' Indian Journal of Agricultural Sciences, 47(7):336-340.

Pasricha, N. S., H. C. Sharma, and N. S. Randhawa. 1972. 'Role of Sulphur in Modern Agriculture,' Indian Farming, 22(4):17-18.

Passmore, R., B. M. Nicol, and M. N. Rao in collaboration with G. H. Beaton and E. M. DeMayer. 1974. Handbook on Human Nutritional Requirements, FAO Food and Nutrition Series No. 4, Food and Agriculture Organization of the United Nations, Rome, Italy.

Pathak, A. N., and S. P. Bhardwaj. 1968. 'Note on the Role of Sulphur in the Nutrition of Berseem (Trifolium alexandrinum L.),' Indian Journal of Agricultural Sciences, 38:1028-1031.

Pathak, A. N., and R. K. Pathak. 1972. 'Effect of Sulphur Component of Some Fertilizers on Groundnut (Arachis hypogaea T.28),' Indian Journal of Agricultural Research, 6:23-26.

Paulino, L. 1980. 'A General View of the World Food Situation,' IN Food Situation and Potential in the Asian and Pacific Region, pp. 1-27, Food and Fertilizer Technology Center Book Series No. 17, Taipei, Taiwan.

Pedraza, L. A., and S. R. Lora. 1974. 'Availability of Sulphur for Plants in Two Soils of the Eastern Plains of Colombia,' Revista, Instituto Colombiano Agropecuario, Colombia, 9(1):77-112.

Perez, A., and D. D. Oelsligle. 1975. 'Comparacion de Differentes Extractantes para Azufre en Suelos de Costa Rica,' Turrialba, 25(3):232-238.

Pinstrup-Andersen, P. 1976. 'Preliminary Estimates of the Contribution of Fertilizer to Cereal Production in Developing Market Economies,' Journal of Economics, 2:169-172.

Potash Institute of North America. Good Acres' Yields Take Up Much Plant Food(pamphlet).

President's Commission on Coal. 1980. Coal Data Book, Government Printing Office, Washington, D.C., U.S.A.

President's Science Advisory Committee. 1967. The World Food Problem, two volumes, Government Printing Office, Washington, D.C., U.S.A.

Pumphrey, F. V., and D. P. Moore. 1965. 'Sulphur and Nitrogen Contet of Alfalfa Herbage During Growth,' Agronomy Journal, 57:237-239.

Probert, M. E., and R. K. Jones. 1977. 'Use of Soil Analysis for Predicting the Response to Sulphur of Pasture Legumes in the Australian Tropics,' Australian Journal of Soil Research, 15(2):137-146.

Ramirez, E., and D. D. Oelsligle. 1978. 'Sulphate Retention in Costa Rican Soils,' Turrialba, 28(2):129-134.

Rathee, O. P., and R. S. Chahal. 1977. 'Effect of Phosphorus and Sulphur Application on the Yield and Chemical Composition of Groundnut (Arachis hypogaea L.) in Ambala Soils,' Journal of Research, Haryana Agricultural University, Hissar, 7(4):173-177.

Reddy, C. S., and B. V. Mehta. 1970a. 'Fractionation of Sulphur in Some Soils of Gujarat to Evolve a Suitable Method for Assessing Available Sulphur Status,' Indian Journal of Agricultural Sciences, 40(1):5-12.

Reddy, C. S., and B. V. Mehta. 1970b. 'Response of Alfalfa (Medicago sativa L.) to Sulphur Application on Loamy-Sand Soils of Anand,' Indian Journal of Agricultural Sciences, 40(5):452-456.

Reddy, C. S., and B. V. Mehta. 1970c. 'Relationship of Carbon, Nitrogen, and Sulphur in Gujarat Soils,' Indian Journal of Agricultural Sciences, 40(7):630-633.

Reisenauer, H. M. 1963a. 'Relative Efficiency of Seed-and-Soil-Applied Molybdenum Fertiliser,' Agronomy Journal, 55:459-460.

Reisenauer, H. M. 1963b. 'The Effect of Sulphur on the Absorption and Utilization of Molybdenum by Peas,' Soil Science Society of America, Proceedings, 27:553-555.

Reisenauer, H. M. 1975. 'Soil Assays for the Recognition of Sulphur Deficiency,' IN Sulphur in Australasian Agriculture, K. D. McLachlan (Ed.), pp. 182-187, Sydney University Press, Sydney, Australia.

Reisenauer, H. M., L. M. Walsh, and R. G. Hoeft. 1973. 'Testing Soils for Sulfur, Boron, Molybdenum, and Chlorine,' IN Soil Testing and Plant Analysis, pp. 173-200, Revised Edition, Soil Science Society of America, Incorporated, Madison, Wisconsin, U.S.A.

Reneau, R. B., Jr. 1981. "Crop Response to Sulphur Application," El Uso del Azufre para Desarrollo y Modernización de la Agricultura en América Latina, 1.er Simposio, Mexico, D.F., México.

Rhue, R. D., and E. J. Kamprath. 1973. 'Leaching Losses of Sulfur During Winter Months When Applied as Gypsum, Elemental S, or Prilled S,' Agronomy Journal, 65(4):603-605.

Richard, L. 1972. "Sulfur Deficiencies in Certain Tropical Crops--A Review of the Conditions Under Which They Occur and Develop," IN Proceedings of International Symposium on Sulfur in Agriculture, Versailles, 3-4.XII.1970, Annales Agronomiques, Numéro hors série, pp. 351-375.

Robinson, E., and R. C. Robbins. 1970. 'Gaseous Sulphur Pollutants From Urban and Natural Sources,' Journal of the Air Pollution Control Association, 20:233-235.

Rowell, A.W.G., and P. M. Grant. 1977. 'Studies on Sulphate Fertilisers for Rhodesian Crops. 2. Effect on Soil Sulphate Status,' Rhodesian Journal of Agricultural Research, 15:33-43.

Ruhal, D. S., and K. V. Paliwal. 1978. 'Status and Distribution of Sulphur in Soils of Rajasthan,' Journal of the Indian Society of Soil Science, 26(4):352-358.

Sachdev, M. S., and P. Chhabra. 1974. 'Transformation of ^{32}S-Labelled Sulphate in Aerobic and Flooded Soil Conditions,' Plant and Soil, 41(2):335-341.

Saggar, S., and G. Dev. 1974. 'Uptake of Sulphur by Different Varieties of Soybean,' Indian Journal of Agricultural Sciences, 44(6):345-349.

Sakai, H. 1980. 'Some Analytical Results of Sulphur-Deficient Plants, Soil and Water,' IN Workshop on Sulphur Nutrition in Rice, December 1978, pp. 35-59, Publication No. 41, Bangladesh Rice Research Institute, Joydebpur, Dhaka, Bangladesh.

Sanchez, P. A. 1976. Properties and Management of Soils in the Tropics, John Wiley and Sons, New York, New York, U.S.A.

Sanchez, P. A., and T. T. Cochrane. 1980. "Soil Constraints in Relation to Major Farming Systems in Tropical America," IN Priorities for Alleviating Soil- Related Constraints to Food Production in the Tropics, pp. 107-139, jointly sponsored and published by the International Rice Research Institute and the New York State College of Agriculture and Life Sciences, Cornell University, in cooperation with the University Consortium of Soils for the Tropics, IRRI, Los Ba:os, Philippines.

Sansum, L. L., and J.B.D. Robinson. 1974. 'Wet Digestion, Manual and Automated Analysis of Total Sulphur in Plant Material,' Commun. Soil Sci. Pl. Anal., 5:365-383.

Saroha, M. S., and H. G. Singh. 1979. 'Effect of Prevention of Iron Chlorosis on the Quality of Sugarcane Grown on Vertisol,' Plant and Soil, 52:467-473.

Schalscha, E. B., C. Estrada, and G. G. Galindo. 1972. 'Sulphur Status of Some Volcanic Ash Derived Soils in Chile,' Agrochimica, 16:1-2, 77-82.

Sen, P. K., and A. Lahiri. 1960. 'Studies on the Nutrition of Oilseed Crops. IV. Effects of Phosphorus and Sulphur on the Uptake of Nitrogen and Growth, Yield, and Oil Content of Sesame (Sesamum indicum L.), Ind. Agric., 4:23-26.

Sharma, R. L., and G. Dev. 1980. 'Interaction of Amide-Nitrogen and Sulphur for Growth and Nutrient Accumulation in Sunflower,' Journal of Nuclear Agricultural Biology, 9(4):146-148.

Shelton, J. E. 1979. Sulfur, Mineral Commodity Profiles, Bureau of Mines, U.S. Department of the Interior, Washington, D.C., U.S.A.

Shelton, J. E., and D. E. Morse. 1983. 'Changing Patterns in Industrial Demand for Sulphur,' quoted in Industrial Minerals, January, p. 41.

Shriniwas, S., K. Kataria, and N. Singh. 1979. 'Note on ^{35}S Studies with Alfalfa to Find Out the Availability of Sulphur in Three Soils of Udaipur Valley,' Indian Journal of Agricultural Sciences, 49(1):63-65.

Shukla, U. C., and A. K. Gheyi. 1971. 'Sulphur Status of Some Rajasthan Soils,' Indian Journal of Agricultural Sciences, 41(3):247-253.

Singh, B., and A. S. Marok. 1980. 'An Appraisal of Quality of Subsoil Irrigation Water of Sunam Block, District Sangrur (Punjab),' Indian Journal of Ecology, 7(1):140-146.

Singh, B. R., A. P. Uriyo, and M. Kilasara. 1979. 'Sorption of Sulphate and Distribution of Total Sulphate and Mineralisable Sulphur in Some Tropical Soil Profiles in Tanzania,' Journal of the Science of Food and Agriculture, 30(1):8-14.

Singh, G. B., and M. K. Moolani. 1970. 'Influence of Sulphur and Nitrogen Levels on Yield and Quality of Raya (Brassica juncea L.),' Bulletin, Indian Society of Soil Science, 8:129-133.

Singh, H. G. 1970. 'Effect of Sulphur in Preventing the Occurrence of Chlorosis in Peas,' Agronomy Journal, 62(6):708-711.

Singh, H. G. 1971. 'Effect of Sulphur on Tissue Composition and Prevention of Chlorosis in Rice Seedling,' Indian Journal of Agronomy, 16:143-148.

Singh, M., and N. Singh. 1977. 'Effect of Sulphur and Selenium on SlphurContaining Amino Acids and Quality of Oil in Raya (Brassica juncea Coss.) in Normal and Sodic Soils,' Indian Journal of Plant Physiology, 20(1):56-62.

Singh, M., and N. Singh. 1978. 'Effect of Sulphur on the Quality of Raya (Brassica juncea Coss.),' Journal of the Indian Society of Soil Science, 26(2):203-207.

Singh, N., and M. Singh. 1980. 'The Effect of Sulphur and Selenium on Nitrogen, Sulphur, Selenium, and Phosphorus in Raya (Brassica juncea Coss.) Grown in Some Soils,' Journal of the Indian Society of Soil Science, 28(4):491-500.

Singh, N. T. 1978. 'Quality of Irrigation Waters in Punjab; – IN Land and Water Resources National Symposium on Land and Water Management in the Indus Basin, Punjab Agricultural University, Ludhiana, Vol. 1:221-229.

Singh, N. T., G. S. Hira, and M. S. Bajwa. 1981. "Use of Amendments in Reclamation of Alkali Soils in India," Agrokemia :s Talajtan Supplementum Tom., 30.158-177.

Singh, N., B. V. Subbiah, and Y. P. Gupta. 1970. 'Effect of Sulphur Fertilization on the Chemical Composition of Groundnut and Mustard,' Indian Journal of Agronomy, 15:24-28.

Sisodia, A. K., N. J. Sawarkar, and M. M. Rai. 1975. 'Effect of Sulphur and Molybdenum on Yield and Nutrient Uptake in Berseem (Trifolium alexandrinum),' Journal of the Indian Society of Soil Science, 23(1-4):96-102.

238

Slack, A. V., and J. O. Hardesty. 1964. 'Status of Superphosphate in Relation to Other Fertilizer Phosphates,' IN Superphosphate: Its History, Chemistry and Manufacture, pp. 315-339, U.S. Department of Agriculture and Tennessee Valley Authority, Government Printing Office, Washington, D.C., U.S.A.

Spedding, C.R.W., J. M. Walsingham, and A. M. Hoxey. 1981. Biological Efficiency in Agriculture, Academic Press, London, England.

Spencer, K. 1966. 'Soil Properties in Relation to the Sulphur and Phosphorus Status of Some Basaltic Soils,' Australian Journal of Soil Research, 4:115-130.

Spencer, K. 1975. 'Sulphur Requirements of Plants,' IN Sulphur in Australasian Agriculture, K. D. McLachlan (Ed.), pp. 98-116, Sydney University Press, Sydney, Australia.

Spencer, K., and N. J. Barrow. 1963. 'A Survey of the Plant Nutrient Status of the Principal Soils of the Northern Tablelands of New South Wales,' CSIRO Aust. Div. Pl. Ind. Tech. Paper No. 19.

Starkey, R. L. 1950. 'Relation of Microorganisms to Transformations of Sulphur in Soils,' Soil Science, 70:55-65.

Stephens, D. 1960. 'Fertilization Experiments with Phosphorus, Nitrogen, and Sulphur in Ghana,' Empire Journal of Experimental Agriculture, 28:151-164.

Stewart, B. A. 1969. 'N:S Ratio: A Guideline to Sulphur Needs,' Sulphur Institute Journal, 5(3):12-15.

Stewart, B. A., and L. K. Porter. 1969. 'Nitrogen-Sulphur Relationships in Wheat (Triticum aestivum L.), Corn (Zea mays) and Beans (Phaseolus vulgaris),' Agronomy Journal, 61(2):267-271.

Stewart, B. A., L. K. Porter, and F. G. Viets, Jr.. 1966. 'Effect of Sulphur Content of Straws on Rates of Decomposition and Plant Growth,' Soil Science Society of America, Proceedings, 30:355-358.

Storey, H. H., and R. Leach. 1933. 'A Sulphur-Deficiency Disease of the Tea Bush,' Annals of Applied Biology, 20:23-56.

Subbarao, A., and A. B. Ghosh. 1981. 'Effect of Intensive Cropping and Fertilizer Use on the Crop Removal of Sulphur and Zinc and Their Availability in Soil,' Fertilizer Research, 2:303-308.

Subbiah, B. V., and N. Singh. 1970. 'Efficiency of Gypsum as a Source of Sulphur to Oil Seed Crops Studied With Radioactive Sulphur and Radioactive Calcium,' Indin Journal of Agricultural Sciences, 40(3):227-234.

−Sulphur Classified as a Macronutrient.− 1978. Sulphur in Agriculture, 2:21.

The Sulphur Institute. 1975. Sulphur: The Essential Plant Nutrient, The Sulphur Institute, Washington, D.C., U.S.A.

Sumbak, J. H., and E. Best. 1976. 'Fertiliser Response With Coconuts in Coastal Papua,' Papua and New Guinea Agricultural Journal, 27(4):93-102.

Swarup, A., and A. B. Ghosh. 1980. 'Changes in the Status of Water Soluble Sulphur and Available Micronutrients in Soil as a Result of Intensive Cropping and Manuring,' Journal of the Indian Society of Soil Science, 28(3):366-370.

Swindale, L. D. 1982. 'Distribution and Use of Arable Soils in the Semi-Arid Tropics,' IN Managing Soil Resources, pp. 67-100, Plenary Session Papers, 12th International Congress of Soil Science, Feb. 8-16, New Delhi, India.

Tabatabai, M. A. 1982. 'Analytical Methods for Sulphur in Soils,' IN Proceedings of the International Sulphur '82 Conference, held in London, November 14-17, 1982, A. I. More (Ed.), Vol. I:391-408, British Sulphur Corporation, Ltd., London, England.

Tergas, L. E. 1977. 'Importance of Sulphur in the Mineral Nutrition of Tropical Forage Legumes,' Turrialba, 27(1):63-70.

Terman, G. L. 1978. Atmospheric Sulfur--The_____ Agronomic Aspects, Tech. Bull. No. 23, The Sulphur Institute, Washington, D.C., U.S.A.

Tilton, J. E. 1977. The Future of Nonfuel Minerals, The Brookings Institution, Washington, D.C., U.S.A.

239

Tisdale, S. L. 1977. Sulphur in Forage Quality and Ruminant Nutrition, Technical Bulletin 22, The Sulphur Institute, Washington, D.C., U.S.A.

Tisdale, S. L., and W. L. Nelson. 1975. Soil Fertility and Fertilizers, 3rd Edition, The Macmillan Co., New York, New York, U.S.A.

Tisdale, S. L., and J. S. Platou. 1981. "The Importance of Sulphur in Tropical Agriculture," El Uso del Azufre para el Desarrollo y Modernización de la Agricultura en América Latina, 1.er Simposio, México, D.F., México.

Troll, C. 1965. 'Seasonal Climate of the Earth,' IN World Maps of Climatology, E. Rodenwalt and H. Justaz (Eds.), pp. 40-53, Springer-Verlag, Berlin.

United Nations Industrial Development Organization (UNIDO). 1978. Second World-Wide Study on the Fertilizer Industry: 1975-2000, UNIDO/ICIS.81, International Center for Industrial Studies, United Nations Industrial Development Organization, Vienna, Austria.

United States Department of Agriculture (USDA). 1954. Diagnosis and Improvement of Saline and Alkali Soils, Agricultural Handbook No. 60, U.S. Salinity Laboratory, Government Printing Office, Washington, D.C.

United States Department of Agriculture (USDA). 1974. The World Food Situation and Prospects to 1985, Foreign Agricultural Economic Report No. 98, Economic Research Service, Washington, D.C., U.S.A.

Valverde, E., E. Bornemisza, and A. Alvarado. 1978. 'Sulphur Availability in Some Soils of the North-Atlantic Region of Costa Rica,' Agronomia Costarricense, 2(2):147-155.

Van Breemen, N., and L. J. Pons. 1978. "Acid Sulfate Soils and Rice," IN Soils and Rice, pp. 739-761, International Rice Research Institute, Los Ba:os, Philippines.

Venkateswarlu, J. 1971. 'Direct and Residual Effects of Sulphates in Crop Production,' IN Proceedings of the International Symposium on Soil Fertility Evaluation, Indian Society of Soil Science, V.1, 921-925, New Delhi.

Venkateswarlu, J., B. V. Subbiah, and R. V. Tamhane. 1969. 'Vertical Distribution of Forms of Sulphur in Selected Rice Soils of India,' Indian Journal of Agricultural Sciences, 39(5):426-431.

Verma, K. S., and I. P. Abrol. 1980. 'Effect of Gypsum and Pyrites on Yield and Chemical Composition of Rice and Wheat Grown in a Highly Sodic Soil,' Indian Journal of Agricultural Sciences, 50(12):935-942.

Virmani, S. M. 1971. 'Comparative Efficiency of Different Methods for Evaluating Available Sulfur in Soils,' Indian Journal of Agricultural Sciences, 41(2):119-125.

Virmani, S. M., and H. C. Gulati. 1971. 'Effect of Sulphur on the Response of Indian Mustard (Brassica juncea L.) to Phosphorus Fertilization,' Indian Journal of Agricultural Sciences, 41(2):143-146.

Vogt, J.B.M. 1966. 'Responses to Sulphur-Fertilization in Northern Rhodesia,' Agrochimica, 10:105-113.

Von Peter, A. 1980. Fertilizer Requirements in Developing Countries, Proceedings No. 188, The Fertiliser Society, London, England.

Walker, D. R., and G. Doornenbal. 1972. 'Soil Sulphate (II) As An Index of Sulphur Available to Legumes,' Canadian Journal of Soil Science, 52:261-266.

Wang, C. H. 1978. 'Sulphur Fertilization of Rice,' Sulphur in Agriculture, 2:13-16.

Wang, C. H., T. H. Liem, and D. S. Mikkelsen. 1976a. Sulfur Deficiency--A Limiting Factor in Rice Production in the Lower Amazon Basin. I. Development of Sulphur Deficiency as a Limiting Factor for Rice Production, IRI Bulletin 47, IRI Research Institute, Inc., New York, New York, U.S.A.

Wang, C. H., T. H. Liem, and D. S. Mikkelsen. 1976b. Sulfur Deficiency--A Limiting Factor in Rice Production in the Lower Amazon Basin. II. Sulfur Requirement for Rice Production, IRI Bulletin 48, IRI Research Institute, Inc., New York, New York, U.S.A.

Watson, K. A. 1964. 'Les Engrais en Nigeria du Nord, Utilization Actuelle et Recommendation D'Emploi,' Sols Africains, 9:21-38.

Wendt, W. B. 1970. 'Responses of Pasture Species in Eastern Uganda to Phosphorus, Sulphur, and Potassium,' East African Agricultural and Forestry Journal, 36(2):211-219.

240

Western Canada Fertiliser Association. 1978. Plant Nutrients Used by Crops. October.

Weterings, K. 1982. The Utilization of Phosphogypsum, Proceedings No. 208, The Fertiliser Society, London, England.

Whitehead, D. C. 1964. 'Soil and Plant Nutrition Aspects of the Sulphur Cycle,' Soils and Fertilizers, 27:1-8.

Williams, C. H. 1967a. 'Nitrogen, Sulphur, and Phosphorus – Their Interactions and Availability,' International Society of Soil Science, Joint Meeting of Commission 2 (Soil Chemistry) and Commission 4 (Soil Fertility and Plant Nutrition), Transactions, pp. 93-111.

Williams, C. H. 1967b. 'Some Factors Affecting the Mineralization of Organic Sulphur in Soils,' Plant and Soil, 26:205-223.

Williams, C. H. 1975. 'The Chemical Nature of Sulphur Compounds in Soils,' IN Sulphur in Australasian Agriculture, K. D. McLachlan (Ed.), pp. 21-30, Sydney University Press, Sydney, Australia.

World Bank. 1982. World Development Report 1982, Oxford University Press, New York, New York, U.S.A.

Wortman, S., and R. W. Cummings, Jr. 1978. To Feed This World: The Challenge and the Strategy, The Johns Hopkins University Press, Baltimore, Maryland, U.S.A.

Yadav, J.S.P. 1982. 'Water Quality and Soil Productivity,' IN Review of Soil Research in India, Part II, pp. 622-634, 12th Int. Congress of Soil Science, New Delhi, India.

Yoshida, S., and M. R. Chaudhry. 1979. 'Sulphur Nutrition of Rice,' Soil Science and Plant Nutrition, 25(1):121-134.

Zake, J.Y.K. 1972. The Effect of Sulphur on the Yield and Quality of Finger Millet in Uganda, Ph.D. diss., Ohio State University, Columbus, Ohio, U.S.A.

Appendixes

List of Appendixes

Appendix I

Nutrient contents of crops[a]

Crop		Yield (mt/ha)	Nutrient content (kg/ha)					
			N	P	K	Ca	Mg	S
Cereals								
Maize	Grain	5.0	115	28	35	2	10	11
	Total	15.0	170	35	175	27	39	19
	Grain	12.5	168	42	53	–	20	17
	Total	25.0	298	55	247	–	73	37
Sorghum	Grain	2.5	40	6	8	5	6	4
	Total	5.0	65	10	48	16	12	7
	Grain	8.9	134	29	28	–	16	25
	Total		280	44	186	–	50	43
Wheat	Grain	3.0	75	15	12	3	9	5
	Total	8.0	125	22	92	16	14	14
	Grain	5.4	106	21	25	–	13	6
	Total		153	26	150	–	26	23
Rice	Grain	3.0	46	8	13	2	5	4
	Total	8.0	84	14	89	21	9	9
	Grain	7.8	86	23	26	–	9	6
	Total		125	30	137	–	16	14
Barley	Grain	5.4	123	20	32	–	9	11
	Total		168	27	139	–	19	22
Protein and oil crops								
Soybeans	Beans	3.0	200	26	57	10	10	6
	Total	9.0	300	40	115	70	35	23
	Beans	4.0	269	24	78	–	19	13
	Total		363	31	132	–	30	28
Sunflowers	Seed	3.9	140	29	36	–	13	7
	Total		197	34	120	–	47	18
Groundnuts	Nuts	3.0	142	15	30	5	10	8
	Total	9.0	323	31	170	118	31	24
Oil palms		24.6	73	12	92	–	21	–
	Total		193	36	249	–	61	–
Coconut	Nuts, husk		35	8	71	1	4	–
	Foliage, total		74	16	113	12	18	–
Field beans	beans	1.0	37	4	22	4	4	10
	Total	3.0	102	9	93	54	18	25
Peas	Peas		25	4	9	4	2	2
	Total		80	8	60	25	8	15
	Peas	2.8	103	12	29	–	7	7
	Total		184	17	98	–	20	11

Crop		Yield (mt/ha)	Nutrient content (kg/ha)					
			N	P	K	Ca	Mg	S
Sugar and starch crops								
Sugarcane	Stalks[b]	103.0	76	14	110	30	29	25
	Stalks	224.0	179	44	311	–	45	60
	Total		403	76	567	–	112	96
Sugar beets	Beets	67.0	140	7	232	–	30	11
	Total		286	20	511	–	89	50
Potatoes	Tubers	40.0	80	5	100	3	3	3
	Total		200	8	220	52	17	11
	Tubers	56.0	194	36	261	–	17	17
	Total		302	44	508	–	57	25
Cassava	Roots	19.0	39	4	32	12	6	2
	Total		113	11	79	62	18	8
	Roots	45.0						
	Total		202	32	286	131	108	15
Sweet potatoes	Roots	33.6	82	19	157	–	9	–
	Total		175	34	290	–	20	–
Horticultural crops								
Onions	Bulb	37.0	66	14	77	4	5	22
	Total		133	22	177	16	18	34
Cabbage	Heads	84.0	140	17	128	–	9	64
	Total		280	31	249	–	36	64
Tomatoes	Fruit	41.0	72	18	130	7	7	9
	Total		84	21	185	31	8	28
Stimulants								
Cacao	Beans	1.0	20	6	30	3	4	–
Coffee	Beans	2.0	33	3	52	7	3	3
	Total		253	19	232	143	33	27
Forage crops								
Grasses		13.0	200	30	200	50	50	20
		25.0	300	70	500	150	100	75
Clover grass		13.4	336	44	335	–	34	34
Alfalfa		22.4	672	58	558	280	59	57
Fruit crops								
Bananas	Fruit	30.0	142	18	365	10	–	–
	Total		627	69	1 390	278	–	–
Papaya	Fruit	2.0	74	8	15	10	11	–
Oranges	Fruit	6 boxes per tree	150	24	240	90	24	15
Fiber crops								
Cotton	Lint	1.68	105	43	49	–	12	8
	Total		201	71	141	–	39	34
Sisal			122	25	216	266	–	–

a. Originally from Malavolta (1979) and English translation from Munson (1982).
b. Programma Nacional De Melhoramento Da Cana-De-Acucar, Relatorio Anual, 1976, Brasil.

245

Appendix II

Sulfur-containing fertilizer materials[a]

Material	Formula	Nitrogen	P₂O₅	K₂O	Sulfur	Other	Sulfur content (lb/short ton)
			Plant Nutrient Content (%)				

Material	Formula	Nitrogen	P$_2$O$_5$	K$_2$O	Sulfur	Other	Sulfur content (lb/short ton)
Aluminum sulfate	Al$_2$SO$_4$·18H$_2$O	0	0	0	14.4	11.4 (Al)	288
Alunite	K$_2$Al$_6$(OH)$_{12}$(SO$_4$)$_4$	0	0	10.5	14.1	17.9 (Al)	282
Ammonia-sulfur solution	NH$_3$+S	74	0	0	10		200
		70.5					280
Ammonium bisulfite	NH$_4$HSO$_3$	14.1	0	0	32.3		646
Ammonium bisulfite solution	NH$_4$HSO$_3$ + H$_2$O	8.5	0	0	17		340
Ammonium nitrate-sulfate	b	30	0	0	5		100
Ammonium phosphate	MAP (crude)	11	48	0	2.2		44
Ammonium phosphate-sulfate	MAP, DAP + (NH$_4$)$_2$SO$_4$	16.5	20.5	0	15.5		310
		13	39	0	7		140
Ammonium polysulfide	NH$_4$S$_x$	20.5	0	0	45		900
Ammonium polysulfide solution	NH$_4$S$_x$	20	0	0	40		800
Ammonium sulfate	(NH$_4$)$_2$SO$_4$	21	0	0	24.2		484
Ammonium sulfate-nitrate	(NH$_4$)$_2$SO$_4$·NH$_4$NO$_3$	26	0	0	12.1		242
Ammonium thiosulfate	(NH$_4$)$_2$S$_2$O$_3$	18.9	0	0	43.3		866
Ammonium thiosulfate solution	(NH$_4$)$_2$S$_2$O$_3$ + H$_2$O	12	0	0	26		520
Apthitalite	(K,Na)$_3$(NaSO$_4$)$_2$	0	0	42.5	15.1		302
Aqua-sulfur solution	NH$_3$+NH$_3$S$_x$+H$_2$O	20	0	0	5		100
Basic slag (Thomas)		0	15.6	0	3		60
Cement flue dust		0	0.6	6	3	3 (Mg)	60
Cobalt sulfate	CoSO$_4$·7H$_2$O	0	0	0	11.4	21 (Co)	228
Copper sulfate	CuSO$_4$·5H$_2$O	0	0	0	12.8	25.5 (Cu)	256

Appendix II. Continued.

Material	Formula	Plant Nutrient Content (%)					Sulfur content (lb/short ton)
		Nitrogen	P_2O_5	K_2O	Sulfur	Other	
Ferrous ammonium sulfate	$Fe(NH_4)_2(SO_4)_2$	6	0	0	16	16 (Fe)	320
Ferrous sulfate	$FeSO_4 \cdot H_2O$	0	0	0	18.8	32.8 (Fe)	376
Ferrous sulfate (copperas)	$FeSO_4 \cdot 7H_2O$	0	0	0	11.5	20 (Fe)	230
Glaubers salt	$Na_2SO_4 \cdot 10H_2O$	0	0	0	10		200
Guano (Peruvian)		0	0	11	1.1		22
Gypsum (anhydrite)	$CaSO_4$	0	0	0	23.5	41.1 (CaO)	470
Gypsum (hydrated)	$CaSO_4 \cdot 2H_2O$	0	0	0	18.6	32.6 (CaO)	372
Gypsum (byproduct)[c]		0	2.5	0	17.2	21.6 (CaO)	344
Gypsum (impure)[d]		0	0	0	13.6	23.8 (CaO)	272
Kainite	$MgSO_4 \cdot KCl \cdot 3H_2O$	0	0	19	12.9	9.7 (Mg)	258
Kalinite	$K_2SO_4 \cdot Al_2(SO_4)_3 \cdot 24H_2O$	0	0	9.9	13.5	5.7 (Al)	270
Kieserite	$MgSO_4 \cdot H_2O$	0	0	0	23	17.5 (Mg)	460
Krugite	$K_2SO_4 \cdot MgSO_4 \cdot 4CaSO_4 \cdot 2H_2O$	0	0	10.7	21.9	2.8 (Mg)	438
Leonite	$K_2SO_4 \cdot MgSO_4 \cdot H_2O$	0	0	25.5	20.5	7.8 (Mg)	410
Lime sulfur (dry)	CaS_x	0	0	0	57	43 (Ca)	1 140
Lime sulfur (solution)	$CaS_5 + Ca_2SO_3 \cdot 5H_2O +$ $CaS_4 + CaSO_3 \cdot 2H_2O$		0	0	23–24	9 (Ca)	480
Magnesium sulfate	$MgSO_4$	0	0	0	30	20 (Mg)	600
Magnesium sulfate (Epsom salt)	$MgSO_4 \cdot 7H_2O$	0	0	0	13	9.8 (Mg)	260
Manganese sulfate	$MnSO_4 \cdot 4H_2O$	0	0	0	14.5	25 (Mn)	290
Polyhalite	$2CaSO_4 \cdot MgSO_4 \cdot K_2SO_4 \cdot 2H_2O$	0	0	15.6	21.2	4.0 (Mg)	424
Potassium sulfate	K_2SO_4	0	0	50	17.6		352
Pyrites	FeS_2	0	0	0	53.5[f]	46.5 (Fe)	1 070
Schoenite (Picromerite)	$K_2SO_4 \cdot MgSO_4 \cdot 6H_2O$	0	0	23.3	15.9	6.0 (Mg)	318
Sodium bisulfate (Nitre cake)	$NaHSO_4$	0	0	0	26.5		530
Sodium sulfate (salt cake)	Na_2SO_4	0	0	0	22.6		452
Sulfate of potash-magnesia (Langbeinite)	$K_2SO_4 \cdot 2MgSO_4$	0	0	22	22	11 (Mg)	440

Appendix II. Continued.

Material	Formula	Plant Nutrient Content (%)					Sulfur content (lb/short ton)
		Nitrogen	P₂O₅	K₂O	Sulfur	Other	
Sulfuric acid (100%)	H_2SO_4	0	0	0	32.7		654
Sulfuric acid (66° Bé = 93%)	H_2SO_4	0	0	0	30.4		608
Sulfuric acid (60° Bé = 77.7%)	H_2SO_4	0	0	0	25.4		508
Sulfuric acid (56° Bé = 71.17%)	H_2SO_4	0	0	0	23.2		465
Sulfur	S	0	0	0	100		2 000
Sulfur dioxide	SO_2	0	0	0	50		1 000
Superphosphate, concentrated	g	0	54	0	1.5		30
Superphosphate, normal	h	0	20	0	13.9		278
Superphosphate, 20% normal, ammoniated	i	4.6	19	0	12.0		240
Superphosphate, triple	j	0	46	0	1.5		30
Superphosphate, triple, ammoniated	k	6.9	42	0	1.4		28
Syngenite	$K_2SO_4 \cdot CaSO_4 \cdot H_2O$	0	0	28.8	19.5	12.2 (Ca)	390
Urea-gypsum	$CaSO_4 \cdot 4CO(NH_2)_2$	17.3	0	0	14.8		296
Urea-sulfur	$CO(NH_2)_2 + S$	40	0	0	10		200
Zinc sulfate	$ZnSO_4 \cdot H_2O$	0	0	0	17.8	36.4 (Zn)	356

a. From Bixby and Beaton (1970).
b. NH_4NO_3 (39%), $3NH_4NO_3 \cdot (NH_4)_2SO_4$ (49%), using an ammoniator-granulator.
c. Average content of byproduct from wet-process phosphoric acid plant.
d. Average purity of agricultural gypsum (73% $CaSO_4 \cdot 2H_2O$).
e. Often has considerable NaCl content, and may have as low as 12% K_2O.
f. Commercial pyrites average 48% – 50% sulfur.
g. Heat-treated triple superphosphate (i.e., not made with superphosphoric acid). Analysis will be that of regular triple superphosphate as affected by the amount of water removed.
h. $Ca(H_2PO_4)_2 \cdot 2H_2O$, 3%; $Ca(H_2PO_4)_2$ anhydrous, 17%; $CaSO_4 \cdot 2H_2O$, 9%; $CaSO_4$ anhydrous, 41%; H_3PO_4, 8%; other 22%.
i. $NH_4H_2PO_4$, 5.5%; $(NH_4)_2SO_4$, 17.5% $Ca_3(PO_4)_2$, 28% $CaCO_3$, 37%, other 12%. (Ammoniated at 6 lbs NH_3 per unit of P_2O_5).
j. $Ca(H_2PO_4)_2 \cdot H_2O$, 63% – 73%; $CaHPO_4$ and other phosphates, 17% – 29%; $CaSO_4 \cdot 2H_2O$, 3% – 6%, other 6% – 12%.
k. $CaHPO_4$, 35%; $NH_4H_2PO_4$, 30%; $(NH_4)_2HPO_4$, 10%; $Ca_3(PO_4)_2$, 8%; $(NH_4)_2SO_4$, 2%, other 15% (ammoniated at 4 lbs of NH_3 per unit of P_2O_5).